Tensions in Diversity

Spaces for Collective Life
in Los Angeles

FELICITY HWEE-HWA CHAN

UNIVERSITY OF TORONTO PRESS
Toronto Buffalo London

© University of Toronto Press 2022
Toronto Buffalo London
utorontopress.com
Printed in the U.S.A.

ISBN 978-1-4875-4512-3 (cloth)
ISBN 978-1-4875-4514-7 (EPUB)
ISBN 978-1-4875-4531-4 (PDF)

Library and Archives Canada Cataloguing in Publication

Title: Tensions in diversity : spaces for collective life in Los Angeles /
    Felicity Hwee-Hwa Chan.
Names: Chan, Felicity Hwee-Hwa, author
Description: Includes bibliographical references and index.
Identifiers: Canadiana (print) 20220267472 | Canadiana (ebook) 20220267618 |
    ISBN 9781487545123 (cloth) | ISBN 9781487545314 (PDF) |
    ISBN 9781487545147 (EPUB)
Subjects: LCSH: Public spaces – California – Los Angeles County. |
    LCSH: Minorities – California – Los Angeles County. | LCSH: Cultural
    pluralism – California – Los Angeles County. | LCSH: City planning –
    Social aspects – California – Los Angeles County. | LCSH: Sociology,
    Urban – California – Los Angeles County. | LCSH: Los Angeles
    County (Calif.) – Ethnic relations.
Classification: LCC HT185 .C43 2022 | DDC 307.7609794/93 – dc23

We wish to acknowledge the land on which the University of Toronto
Press operates. This land is the traditional territory of the Wendat, the
Anishnaabeg, the Haudenosaunee, the Métis, and the Mississaugas of the
Credit First Nation.

University of Toronto Press acknowledges the financial support of the
Government of Canada, the Canada Council for the Arts, and the Ontario
Arts Council, an agency of the Government of Ontario, for its publishing
activities.

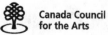

Contents

Figures and Tables

Figures

Tables

Acknowledgments

I would never have imagined writing a book, and certainly not one about a complex place like Los Angeles. This book has survived multiple relocations across the oceans, and I am thankful that I can finally pen the overdue acknowledgments for all the support that I have received to make it a reality.

Foremost, I want to thank the participants of the study from the neighbourhoods and city halls of the Los Angeles metropolitan area who have helped me understand their experiences of living in the city. Through them, I have also been able to make better sense of mine.

I am grateful for the editors at the University of Toronto Press who believed in the project and patiently worked with me over the years to refine the manuscript with the help of anonymous reviewers. Thank you, Jodi Lewchuk, Doug Hildebrand, Perrin Lindelauf, Leah Connor, Stephanie Mazza, and the publishing team at the press for your support.

This book would not have come about without the good supervision of my PhD research study by Tridib Banerjee, Alison Renteln, Dowell Myers, and Janet Hoskins at the University of Southern California. For your open-mindedness towards interdisciplinary research and for your critical insights of the project, I am deeply grateful. My heartfelt thanks to Lisa Schweitzer too: without your prompting, I would never have thought of writing a book based on my research of Los Angeles!

One thing led to another. The postdoctoral fellowship at the Max Planck Institute for the Study of Religious and Ethnic Diversity in Göttingen, Germany significantly widened my horizons on the study of diversity outside the United States. Thank you, Steve Vertovec and Karen Schönwälder, for believing that my project would make a good book and for the opportunity to meet so many like-minded researchers. At the Institute, I had countless insightful and memorable conversations. Thank you, Jörg Hüttermann, Annelies Kusters, Maria Schiller,

Alex Street, Elena Gadjanova, Zeynep Yanasmayan-Wegele, and Julia Martínez-Ariño, for the intellectual camaraderie and for your friendship.

At the Nanyang Technological University Singapore, I want to thank the students in my course *What Is a City?* for the engaging classroom discussions that have further refined the contents of this book. Special thanks to my students, Hui Lee Low and Haoyu Zhao, who have helped me with aspects of the manuscript preparation when I became busy with the travails of the tenure track.

I would be remiss if I did not thank my friends, who have run this book marathon with me from its genesis in 2013 till 2022. Thank you for your faithful prayers and loving encouragement across time zones. Noelle Nienow Soi, Wouter and Eileen Waalewijn, Juliana Zhu, Cheong Sin Kim and Sanghee Park, Michael Lin, Elena Maggioni and Angelo Messi, Stephan and Theres Klühs, Bhramara Tirupati, Jeff and Livia Blackburne, Michelle Busick, Linda Murphy, Traci Black, Gene and Dotty Belknap, Lydia Kung, and Lye Heng and Susan Choo – thank you!

There are two people whose friendship has been particularly important to the fruition of this book. To my dearest friend, Yanzhen Lui, who came to my rescue at a critical juncture and generously invested her time to refine the text: thank you for being the faithful Sam to Frodo. To my dearest soulmate, Ji-Jon Sit: thank you for staying in this project with me, even when I had given up on many occasions.

To Mom, Dad, and Jeff: thank you for helping me put the successes and failures of academic life in perspective.

Thank you, Abba Father, for your blessing of strength and divine encounters with all these individuals so that this book project can be realized in your time.

TENSIONS IN DIVERSITY

1 Introduction: The Promise and Peril of Los Angeles

Entering Los Angeles is like walking into a giant kaleidoscope of multiple shapes, colours, and angles. A metropolis of ten million people coming from different lands and speaking different languages, Los Angeles pulsates with cultural diversity, with life and strife each day. It is a place that fascinates and stimulates the senses of its inhabitants. It is also a city where extreme poverty and runaway wealth are found side by side, where cultural fusion and fission unfold concurrently, where lovers and haters of the city search for niches to find their footing in a sprawling landscape of endless freeways. In this city, the celebration of diversity coexists with segregation of all kinds – ethnicity, race, nationality, income, class, lifestyle, age, sexual orientation, and more. It is a space of perceived freedom for many, yet it is hemmed in by the geography of the Pacific Ocean, the daily traffic gridlocks on its freeways and the threat of fires, earthquakes, and landslides. Los Angeles is also a space of perceived *un*freedoms – real and imagined – where one cannot and will not go everywhere, limited by fear, prejudices, and poverty. These juxtapositions of social, cultural, and economic differences make Los Angeles a multilayered landscape of diversity that is rich with questions about coexistence.

Los Angeles has witnessed several violent riots throughout its history. The episodes of civil unrest have continually made clear that peaceful coexistence among different groups is not a given, particularly in places where multiple fault lines of socio-economic inequality intersect and compound with racism and xenophobia. From the 1871 Chinese Massacre, to the 1965 Watts riot and the 1992 civil unrest, the violent conflict between different racial and ethnic groups repetitively undermines the notion that America is post-racial. More than that, it challenges us to confront the depth of our human inability to achieve convivial coexistence amid differences. These violent events often leave an immediate trail of soul-searching questions about the human condition in their wake. Unfortunately, the time that heals also has the inadvertent capacity to dull our commitment to vigilantly countering the friction and conflicts that arise from differences.

We are now living in the "diversity explosion" of twenty-first-century America that demographer William Frey (2015) wrote about. Post-1965 immigration and globalization have introduced new foreign immigrants from many more different countries, particularly Central America and Asia, making patterns of divisions ever more complex in Los Angeles. Ethnic groups are now further diversified by national origins and class, blurring the Black-White racial line with new racial categories of Asian Americans and Hispanic Americans of different classes.[1] Sociocultural diversity adds to the richness of social life of a city but also increases its social complexity and exposure to fragmentation.

This book begins from the conundrum of diversity in collective life with an eye to discovering what makes life together possible through the lens of Los Angeles's multi-ethnic and multinational locales. I will avoid speaking in the abstract about diversity as a phenomenon of advantage or disadvantage to the economy and society. Instead, this book aims to bring the voices and feelings of those living in diverse locales to the conversation on what diversity is and how it is experienced socially and spatially through the use of textual and visual mapping analysis. All conversations recorded in this book are with real persons; however, the names of the participants have been changed to protect their privacy.

The book draws on my interviews with 140 residents of native and immigrant origin, community organizers, and municipal officers about their social lives and routines in three ethnically mixed and multinational locales of Los Angeles County, namely the City of San Marino, the Central Long Beach area in the City of Long Beach, and the Mid-Wilshire area in Los Angeles, between 2010 and 2012. During these interviews, participants were also asked to map their perception of socio-spatial diversity in their neighbourhoods and to complete a survey about their views on intercultural understanding and contact in public places.[2] To enrich my understanding of the social life in these diverse locales, I also participated in community events, such as a cultural festival, a parade, several council meetings, and planning committee meetings to observe how public life was organized and undertaken. I will elaborate more on the research design and methodological considerations of the study in chapter 2.

Specifically, the narrative of this book is weaved together by discussion of these questions: How is diversity experienced and conceived by those living in diversity? What are the tensions of diversity, and are there common patterns in different contexts of diversity? How do income inequalities commingle with social and cultural differences in neighbourhood spaces? How and where do neighbours from different ethnicities and nationalities routinely interact? Where are the possibilities for intercultural learning? What kinds of local public spaces might be most productive for growing an individual's capability and capacity to live convivially in diversity?

In essence, the book presents how residents of different ethnicities and nationalities from neighbourhoods of different socio-economic levels reconcile everyday coexistence, and what can be done differently in the way we build our public places for sociocultural diversity to flourish in our cities. Thus, it should be self-evident that underlying the empirical research content of this book is an unapologetically normative concern for socially productive collective life in the presence of divisive forces in society.

"Diversity Explosion": Race, Ethnicity, Nationality, Class, and Income

Demographer William Frey (2015) wrote that America is undergoing a "diversity explosion" that will remake the country. In 2011, the United States reached a historical turning point when more minority babies than White babies were born in a year. However, the United States is not the only country undergoing rapid diversification. According to United Nations Population Division's International Migration Report 2019, the number of people living outside their country of birth as immigrants, foreign workers, and refugees between 1990 and 2019 rose by 78 per cent or 119 million to 272 million.[3] By 2020, there were 280 million international migrants, with still many more residing outside their city and region of birth.[4] This number is increasing rapidly as more people are fleeing war and persecution. The voluntary and involuntary movement and resettlement of people across the world has created new landscapes of diversity, but it has also stoked a cacophony of anti-immigrant sentiments and even open conflicts over different practices and values.

Post-1965 immigration has brought many foreign immigrants from Central America and Asia to Los Angeles. Thus, ethnicity must be understood in the context of diverse national origins. Latino immigrants include Mexicans, Salvadorans, and Guatemalans. In fact, it has been repeatedly pointed out that the "Hispanic" ethnic category is among the most racially diverse, with a multiplicity of national origins and socio-economic statuses. The same could be said of the Asian immigrants that are a multi-ethnic and multinational group made up of Filipinos, Koreans, Guamanians, Japanese, Samoans, Armenians, Iranians, and Indians, in addition to the Chinese who come from different parts of East Asia (Taiwan, Hong Kong, China) and Southeast Asians from Thailand, Myanmar, Cambodia, and Vietnam.[5] According to Waldinger and Bozorgmehr (1996, 16), the new immigrant diversity in Los Angeles was not only multi-origin but also characterized by widely ranging skill and education levels.[6]

In the three locales of Los Angeles presented in this book, their demographic diversity reflected the composition of the "diversity explosion" experienced in America interacting with socio-economic inequalities to produce complicated outcomes. They were selected with the aim to study how sociocultural diversity manifests in places of different household income level (a proxy for class), given the critical influence that income has on determining the social conditions of everyday life in Los Angeles. In addition, these three diverse locales had also experienced intergroup tensions over the last few decades that made them meaningful sites for research to uncover patterns and conditions of discord. Further, studying these settings with different ethnic and nationality mixes within the common geography of metropolitan Los Angeles promised to offer valuable insights of how different compositions of sociocultural diversity could produce similar or different experiences of coexistence. See figure 1.1 for the locations of these three locales with respect to the geography of Los Angeles County.

The choice of ethnicity and nationality as qualitative proxies of diversity does not assume that they are primordial characteristics of diversity. Rather, the interest here is to understand how ethnicity and nationality as "ways of seeing the world" among many others may structure social interaction and the experience of living in diversity (Brubaker, Loveman, and Stamatov 2004, 47). This book recognizes that ethnicity (e.g., cultural heritage, language, religion) and nationality (e.g., citizenship, political beliefs, national culture) are themselves conceptually and perceptually overlapping ways of seeing everyday life. Further, they are also frequently intertwined with the simpler phenomic and visible aspects of racial categories in how people identify themselves and each other in diverse contexts that are socially complex and psychologically taxing to navigate. Thus, race, ethnicity, and nationality are nested social categories that are present, real, and important in the negotiation of diversity. Depending on the life history and circumstances of the individual, one category takes predominance over another, and this can change over time, as Omi and Winant (2015, viii) remind us: race is "unstable, flexible, and subject to constant conflict and reinvention."

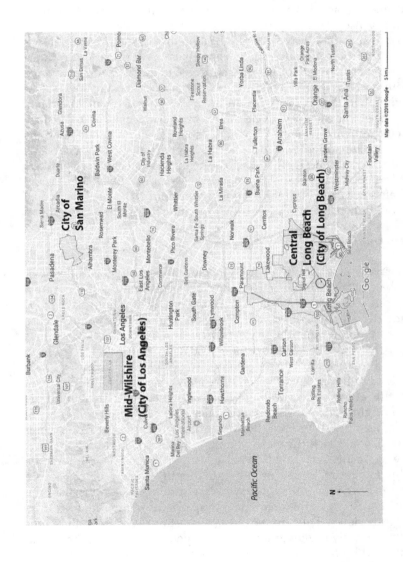

Figure 1.1. Map showing the locations of the three study areas in Los Angeles County. Map is not to scale.

Adapted from Google Street Maps by author.

When the participants were asked how they identify themselves, many first-generation and some second-generation immigrants first mentioned their ethnicity and national origin before adding their current identity as an American. A handful of the second-generation immigrants also identified themselves racially as Asian Americans or Latinos but specified the ethnicity or nationality of their parents. Non-immigrant participants identified themselves predominantly in racial categories such as White or Black Americans. In an interview conversation with an elderly man residing in Central Long Beach, he insisted with pride that he was Black American, and not an African American. To him, the ethnic category of African American was inaccurate because it was culturally irrelevant as he did not share any immediate relations with Africa. He added that if the racial category of White Americans was valid, why should there not also be an equivalent category of Black Americans? For this compelling reason, I will use "Black American" as the reference category in this book rather than "African American," even though the US Census treats them as interchangeable categories.

Locales of Diversity in Los Angeles

San Marino is a small city of high-income households with approximately half of its 13,000 residents having recent origins in East Asia and the other half being White Americans, many of whom have lived there for decades. Wealthy, well-educated, and professional Chinese immigrants predominantly from Taiwan, but also Hong Kong and China, arrived in San Marino in the 1980s and continued to grow in numbers. By 2013, the Chinese population, including many second-generation immigrants, had become the ethnic majority in the city even though this ethnic group was not homogeneous given their multiple national origins. The strained political ties between China and Taiwan, as well as cultural differences between the early cohort of Hong Kong immigrants and recent Chinese immigrants, shaped and divided how Chinese residents identified themselves. For example, while more second-generation Chinese immigrants identified themselves racially as "Asian American," the first-generation immigrants emphasized the specific importance of their national group origins in their identity during the interviews. Thus, in San Marino, nationality rather than ethnicity was socially salient among the first-generation immigrants who were divided by their identities of "Chinese American" or "American Chinese," and "Taiwanese American."

In deep contrast, Central Long Beach is a dense and low-income neighbourhood in the City of Long Beach with an annual median household income that is less than 60 per cent of Los Angeles County's median level in 2010 and 2020.[7] Its residents have lower education attainment, and many do not speak English well. Its population includes first-generation immigrants from Mexico with documented or undocumented statuses, second-generation Mexican Americans, Cambodian refugees who arrived in the 1970s and their American-born children, Vietnamese, Filipinos, and Black American residents who have lived in the neighbourhood for several generations since the 1950s. In 2001, an effort by a few Cambodian businessmen to establish a "Cambodia Town" triggered unease in the multi-ethnic and multinational neighbourhood. It also further fragmented the Cambodian population into pre- versus post-genocide groups. A second division was formed between the more well-to-do Cambodians who had moved out of the poor neighbourhood (but kept their businesses in Central Long Beach) and the poorer Cambodian refugees who could not afford to relocate. In Central Long Beach, there was also no homogenous Asian identity or formation of pan-ethnicity identity because of the tense political history between the Cambodian and Vietnamese people.[8]

Mid-Wilshire in the City of Los Angeles is a dynamic mixed-income zone in midtown with households of very high-, middle- and low-income levels. For many, this area epitomizes the "salad bowl" or "melting pot" character of Los Angeles with over a hundred nationalities residing here. Its residents include those who have recently arrived and long-time residents of Los Angeles. It is home to multiple generations of immigrants from different parts of Guatemala, El Salvador, Mexico, Bangladesh, North Korea and South Korea, China, the Philippines, as well as White Americans and Hassidic Jews. The locale is also socio-economically and physically diverse with densely built rented apartment dwellings, luxurious condominiums, and large single-family homes.

In comparison to the other two locales, I met residents working in a diversity of employment in Mid-Wilshire. They included a first-generation Mexican American who works as a chef in an Italian restaurant and is raising a family in Mid-Wilshire, a long-time White American artist resident who lives alone, a Korean American community organizer, international students from the Philippines and South Korea, first-generation Filipino and Korean immigrant accounting and nursing professionals, a Black American hair stylist, and second-generation Mexican and Guatemalan Americans attending local colleges. Like San Marino and Central Long Beach, intragroup divisions in Mid-Wilshire were evident. In this case, it was within the Latino group. The hostility that many Central American immigrants, namely the first-generation Guatemalans and Salvadorans, experienced in Mexico during their northward migration to the United States motivated many of them and their second-generation children to actively distance themselves from Mexicans by using their parents' place of origins as part of their identity: for example, "Honduran-Black" and "Salvadoran." For those with mixed Mexican heritage like seventeen-year-old Luciana Garcia, she emphasized the mix in her parents' nationalities when asked how she would identify herself. "More Mexican than Guatemalan," she said, explaining that while her everyday life was surrounded by Mexican culture and relatives, a part of her was Guatemalan. Among the Korean participants, I observed a recurrent identity differentiation made by the Korean participants between the 1.5 generation Korean Americans (born in Korea but raised in the United States) and the first-generation Korean immigrants. The first-generation Korean immigrants saw themselves as very different from the 1.5 generation Korean Americans, who, from their point of view, were assimilated Americans.

Therefore, in these three locales, ethnicity and nationality among immigrants were not particularly internally coherent categories but were subdivided by generational differences, political animosities, and class divides that echoed those in the countries where the immigrants came from. In this book, I tried to avoid reifying standard census ethnic and nationality categories as much as possible; instead I used the identifications that people gave themselves and others.[9] Overall, nationality was used more often than ethnicity among the participants as means of self-identification, indicating that in a highly diverse immigrant gateway metropolis like Los Angeles, the sociocultural differentiation among urban residents is experienced at a much finer and more complex grain than any single analytical category of race, ethnicity, nationality, or class alone can describe. In chapter 3, I will describe and discuss in greater detail how residents in each of the three differently diverse locales in Los Angeles perceive, conceive, and experience differences among themselves spatially and socially. See appendix 1 for detailed demographic information of each location and appendix 2 for the list of participants in each location.

Cities of Diversity: Nested Social Complex

From the earliest times, cities have served as multifunctional nodes of refuge, settlement, commerce and exchange, political control, and strife. As the German Middle Ages saying goes, *Stadtluft macht frei* (translated: urban air makes one free): cities around the world have a greater capacity than villages to accommodate differences, enabling them to thrive. The social space of a city is commonly associated with openness and anonymity that can provide both thriving and alienating conditions. A city is also a spatial concentration of social, economic, and political resources. However, the social and cultural diversity that gives cities their dynamism has also the propensity for divisions.

Cities are thus paradoxical places, where close-mindedness and open-mindedness coexist with tendencies to ignite fiery sparks among groups because of clashing values but also present an abundance of opportunities for creative fusion. In globalizing cities that are open to high inflow and outflow of people of different origins and statuses, rapid demographic diversification is unfolding to produce a condition of what anthropologist Steven Vertovec (2007) described as "superdiversity." Demographic diversification thus refers to a process in which standard demographic characteristics like ethnicity, nationality, age, and gender have internally variegated in and of themselves, and combined with other variables – legal status, employment, length of stay, etc. – in different ways, as an outcome of an enhanced global flow of people.

Large, dense, and diverse, a city excites but also overwhelms its inhabitants as a giant contact zone of social and cultural differences. As Mary Louise Pratt (1991, 34 and 39) described, contact zones are "social places where cultures meet, clash and grapple with each other," but they are also places where "exhilarating moments of wonder and revelation, mutual understanding, and new wisdom" are generated. How individuals respond to the stimuli of a large urban complex has been the subject of decades of sociological and psychological research.[10] Sociologist Lyn Lofland (1973, 22) found that to navigate the fear and anxiety in the "world of strangers," individuals had to learn to give order to their encounters in the city by the appearance and spatial location of those they meet. Individuals developed categoric knowledge (e.g., age, gender) via visual and/or verbal information about strangers in the public realm, according to Lofland (1973). This tacit knowledge guides an individual to behave appropriately in the public space among strangers.

In a later study, Lofland (1998, 52–9) developed categories of person-to-person contact in the urban public realm that she termed the "relational web" in public space. These relations are described in ascending order of intimacy as follows: a) *fleeting* relationships referring to interaction between complete strangers; b) *routinized* relationships that refer to the brief "the interaction-as-learned-routine" between categorically known others; c) *quasi-primary* relationships that are encounters lasting between "a few minutes to several hours between strangers or between those who are categorically known to one another"; and d) *intimate-secondary* relationships that are "emotionally infused" and "relatively long-lasting" over a span of weeks to years.

According to Allen and Turner (1996, 25–6), who have analysed the demographic changes from 1970s to 1990s in Los Angeles, the diversification of the city has not in fact meant less segregation, but more. "Asians and Latinos are substantially more segregated from Whites in Los Angeles County than in the average large metropolitan area," such that "Los Angeles is certainly closer to being a mosaic of geographically separate ethnic communities than it is to being a residential melting pot." Even in the case of Mid-Wilshire, where cultural diversity was visible and audible at every street corner because its residents hailed from over a hundred nations, residential segregation in voluntary and involuntary kinds existed alongside the "prismatic" diversity that also characterized Los Angeles (Bobo et al. 2000).

Consider the exchange I had with one of Los Angeles's Asian American residents, which exemplifies the observations of Lofland (1973) and Allen and Turner (1996): Liz Joo and I had known each other for several years by the time we sat down for a conversation about her experiences of living in the Mid-Wilshire area. Liz lived in a private condominium along a street of condominiums sandwiched between streets of big single-family homes with well-kept lawns and streets of densely populated multi-family rented dwellings. According to her description of her neighbourhood during the interview, she lived on the threshold of two very different environments. Her neighbours were composed of lower-income Latinos in multi-family rented dwellings on one side, and mostly high-income Whites and some Asians living in single-family homes on the other. In the diversity of Mid-Wilshire, one finds ethnicity interacting with income to produce visible and severe social and economic inequality in the residential landscape.

MYSELF: Is there much diversity amongst the people living around here?
LIZ: There actually is. I would say on my street there are Whites, Koreans, and Mexicans ... probably the next street over is probably Mexicans. I actually don't walk that way ever. Not even once. It feels dangerous actually. One street east of me. I don't go east of here – of where I live. I never do.
MYSELF: Why?
LIZ: Because it is not safe. There are actually like police reports before that there are assaults. We have gotten it before. People have break-ins. You can tell because it is all apartments. My street is condo ... and starting from this street it is like all multimillion-dollar homes. And I am sure it is all White. So it is kind of diverse in this square because you see Whites, you see some Asians.

MYSELF: Would you mark it down [*on the map*] for me?

LIZ: White is all of Hancock Park, right? I don't know where Hancock Park is. I just know on my street there are Koreans.

MYSELF: How do you tell that it is predominantly White at Hancock Park?

LIZ: 100 per cent. I have never seen anybody else. From the house? No, [you can't tell]. Living there? No. If you are walking there, yeah. But actually coming out from the house, it is all White. If you see a non-White, coz it is the maid. They are walking the dog. Or they are like, you can tell, they are the maids.

As Liz's experience illustrates, nested categories of ethnicity, nationality, income, and class have shaped her daily practices, experiences, and imaginations. Diverse environments present themselves as socially and physically complex to negotiate, and thus categoric knowing among neighbours is common. The resulting complexity has led to an invidious everyday environment where daily spatial practices further reinforce both positive and negative stereotypes that clearly have the potential to entrench any existing prejudice. In chapters 4 and 5, I will present an analysis of the nature of tensions between groups and discuss possibilities for the formation of local belongings given the social and spatial tensions in diversity.

Studying Diversity as a Socio-Spatial Phenomenon

How space is used, occupied, imagined, and experienced is increasingly integral to the understanding of local and international human relations. The reterritorialization of local space following contours of group identities in globalizing cities of immigration (Gupta and Ferguson 1997), accompanied by an "accelerated global de-bordering and re-bordering" of nations geopolitically (Chen 2005, 4–6), underscore the salience of space and spatial thinking in the formation of sociocultural diversity. In fact, as Rumford's (2014, 3) theorizing of the concept of "cosmopolitan borders" illustrated, the socio-spatial lens is an important means to understand intergroup relations as people engage in the "making, shifting and removing" of borders between one another.

To study diversity as a socio-spatial phenomenon, I have used cognitive mapping complemented by personal interviews and participant observations as this mixed method enables simultaneous access to the spatial practices and experiences of individuals living in socioculturally diverse locales. The mixed method foregrounds urban space as an active force in shaping coexistence. Space is not merely a container of actions

but a medium that can shape social relations. This contrasts with sociological studies about intergroup relations in which urban space remains
mostly as a passive backdrop upon which negotiations between people
unfold.

Specifically, I have used the cognitive mapping method to document
the spatial practices of residents and their experiences of diversity in the
three locales. Participants were asked to map their routine spaces, the
locations of intercultural encounters, and their perception of boundaries and territories in their neighbourhoods. Cognitive mapping is valuable as a research tool because of its ability to go beyond the "visually
prominent" to access also "things which are important for historical,
economic, political, and other reasons" (Stea 1974, 161). It has an element of democracy and ground-up participation in place-making.

The mapping method was devised by Kevin Lynch ([1960] 1998 and
[1985] 1996) in a project with György Kepes to understand how the
form of urban environments is perceived by the users of the city.[11] Thus,
the cognitive mapping method enables the visualization of urban space
of an individual that otherwise would be invisible. According to recent
findings in neuroscience, cognitive mapping taps into the hippocampus, which functions not only to store cells and for spatial learning, but
also a storehouse of memories of experiences.[12]

To enrich these maps, I used semi-structured interviews to understand how participants conceptualize diversity, what their experiences of social interaction within the locales were, and what kinds of
intercultural encounters in diversity they had. These interviews about
the residents' thoughts, experiences, and practices of intergroup interaction offer insights into how practices and experiences influenced
mental conceptions of coexistence and vice versa. Combining cognitive mapping with interviews and participant observation enables
studying the interaction of physical space with social relations, i.e.,
social space.

In this book, I make use of Henri Lefebvre's ([1974] 1991) theoretical
conceptualization of social space as produced by an ongoing dialectical interaction among the triad of spatial practices of individuals (perceived space), conceived representations of space by experts (conceived
space), and lived symbolic experiences of the individual (lived space) to
discuss the data collected via cognitive maps and interviews. Lefebvre's
conceptualization of social space is dynamic and valuable as a framework because it deconstructs the complexity of multilayered social
space produced by individuals, groups, and institutions without losing
an emphasis on the dialectical interaction between its elements (Schmid

2008). However, Lefebvre has left the specific workings of the dialectic, whether of conflict or of alliance, open for interpretation, which this book will take the liberty to interpret.

Overlapping the information from combined sources using a Lynch and Lefebvre framework gives insights to how the perceived, conceived, and lived spaces of Lefebvre's triad interconnect in these diverse locales. The reader then shall be the judge of whether urban space is an influential medium in shaping the negotiation of coexistence, as I have claimed.

Lynch's cognitive mapping method enables the scaling up of individual data points for collective analysis of the perceptions of different individuals, and this pairs well with Lefebvre's conceptualization that approaches social space production as a complex of multiple actors. Thus, the Lynch-Lefebvre modus operandi is particularly valuable to study diverse locales where multiple group relations are negotiated through the use and occupation of the neighbourhood as a space of habitation. Without Lynch's method, Lefebvre's theoretical conceptualization of space production would be difficult to test and operationalize. Conversely, without Lefebvre's overarching theoretical framework of space as a reified object of production within a modern society, Lynch's method would lack a theoretical context to ground it and enable discussion of the process of interconnecting the mental and the physical space of habitation. The combination of the methods of mapping, interviews, and participant observation, enabled by adopting the Lynch-Lefebvre framework, is akin to the "thick mapping" method, referred to by Presner, Shepard, and Kawano (2014, 17–18) as the following:

> The processes of collecting, aggregating, and visualizing ever more layers of geographic or place-specific data ... Thickness means extensibility and polyvocality: diachronic and synchronic, temporally layered, and polyvalent ways of authoring, knowing, and making meaning. Not unlike the notion of "thick description" made famous by anthropologist Clifford Geertz, thickness connotes a kind of cultural analysis trained on the political, economic, linguistic, social, and other stratificatory and contextual realities in which human beings act and create.

The "thick mapping" method offers the means to understand and visualize the nested social complex of diverse locales spatially. Further, when this form of mapping is combined with the comparative analysis of three locales, the method provides the possibility of sketching the contours of a diverse public realm – a subject matter of chapters 6 and 7.

More Quality Contact, Not Less

Social psychologist Gordon Allport ([1954] 1979), in his study to understand prejudice formation in a quickly diversifying post-Second World
War United States, proposed more contact between different groups to
reduce prejudice. Popularly known as the "contact hypothesis," Allport
([1954] 1979, 281) underscored that the conditions and opportunities
of contact were extremely important to the quality of contact made in
environments of diversity:

> Prejudice (unless deeply rooted in the character structure of the individ-
> ual) may be reduced by equal status contact between majority and minor-
> ity groups in the pursuit of common goals. The effect is greatly enhanced
> if this contact is sanctioned by institutional supports (i.e., by law, custom,
> or local atmosphere), and provided it is of a sort that leads to the percep-
> tion of common interests and common humanity between members of the
> two groups.

Allport's hypothesis was tested by later generations of social psychologists like Pettigrew and Tropp (2006, 767) who conducted a meta-
analysis of 515 studies and found that intergroup contact under the right
conditions could reduce "feelings of threat and anxiety about future
cross-group interactions." Subsequent studies by Pettigrew and Tropp
(2011) continued to show evidence that friendship between individuals
from different groups promoted positive and high-quality experiences
between their groups that influenced and promoted overall better intergroup well-being. In another meta-analysis study, Christ et al. (2014,
5) further reaffirmed that the availability of intergroup contact in an
environment was essential to reducing prejudice, drawing on research
done across Europe and Africa.

While encounters between individuals of different social and cultural
groups are expected through co-presence in densely populated areas
of social and cultural diversity, the type and depth of personal contact
are variable depending on the opportunities and conditions of contact
that avail an individual. Contact in public space can range from fleeting
unimpressionable encounters to intimate-secondary relations through
which people form emotional attachments over time following Lofland's
(1998) schema outlined above. In addition, as Valentine (2008) points
out, proximity may not always lead to more and better contact opportunities. Instead, proximity creates occasions for comparison between
groups that can exacerbate the perception of socio-economic inequality
and feelings of threats arising from cultural difference. Valentine (2008)

also argues that the toleration of difference may in fact reflect unequal power relations between the agent who is tolerant and hence powerful, and the "weaker" and "less desired" other who is being tolerated. This form of civility via toleration of those in proximity cannot be considered as a mutual respect of difference, further emphasizing that the quality of intergroup contact is judged not by its frequency alone but the sincerity of its exchange and connection as ascertained by the parties involved.[13]

Another form of civility upon encounter is a civility to normalize social and cultural differences by making differences "commonplace" (Wessendorf 2014) and unsurprising. In this way an attitude of indifference to difference, or what Tonkiss (2003) identifies as "ethics of indifference," emerges. Indifference can be freeing as we unlock others from otherness, and as Richard Sennett (2018, 299) pointed out, "Unchained from anthropology, they can then open up to those unlike themselves living in the same place." The "civility of indifference," borrowing from Bailey's (1996) concept, is also a pragmatic coping mechanism to maintain social peace and cooperation in the presence of differences by downplaying and demonstrating an "unreflective unconcern" about difference. This kind of blasé civility can also be a response to the stimulus overload of complex situations presented in urban life, resulting in shutting down mentally and emotionally, according to sociologist Georg Simmel ([1903] 2005) and psychologist Stanley Milgram (1970). Milgram (1970) explained that this stimulus overload would lead individuals to adapt their behaviour by giving less time to each piece of information, prioritizing only the important inputs for retention, shifting responsibilities to the other parties, and preventing further inputs from entering. These social psychological adaptations have implications for the development of collective life in cities. This is not unlike what Gordon Allport ([1954] 1979) suggested as the cause of individuals developing social categories to simplify differences in social complexity.

As Liz Joo's experience illustrated, daily contact in one of the world's most socioculturally diverse places like Los Angeles's Mid-Wilshire area was mostly of a passive nature, composing of largely fleeting encounters between individuals. Quality contact in diversity was not a given. In fact, social life in the diversity present in Los Angeles seemed to manifest measures of indifference to difference mixed with an unease of the proximity of socio-economic and cultural differences. Liz's "hunker down" experience of living in diversity was apparently not exceptional in the United States. In a large-scale study of forty-one neighbourhoods in the United States, Harvard sociologist Robert Putnam (2007) found that in areas of high ethnic diversity levels, inter-racial trust was lower, and the social life reflected a "hunker down" mentality at least in the short-term.

Similarly, findings about the 2001 race riots in northern British towns by Ted Cantle (2005) underscored that coexistence of the Whites and Asians was pervasively "parallel" with little daily interaction or a need for it. In these communities, social and spatial segregation were apparent. Reflecting on 2001 race riots in Britain that were caused by the commingling of ethnic divides with factors including socio-economic deprivation, segregation, and new youth policies, geographer Ash Amin (2002) argued that the multicultural policies practiced in Britain maintained and reinforced cultural differences, rather than permitting and encouraging meaningful intercourse between groups.

Intercultural Learning

Multiculturalism in Britain (also in Canada and Australia) was adopted post-Second World War to address the growing presence of new immigrant cultural groups who were not assimilating, whereas in the United States, the rubric of multiculturalism was formed by multiple political and ideological exigencies in those years, namely the civil rights movement in the early 1960s, and the backlash against the ideology of mass society and nation-building aspirations.[14] One of the greatest criticisms levelled against the practice of multiculturalism is its tendency to reify cultures and "fossilize" cultural differences in codified laws and regulations that can produce schisms between cultures and people.[15] Thus, in societies governed by multicultural goals, encounters between socially and culturally different individuals can remain a demonstration of the civility of indifference or even tolerance that can accentuate negative stereotypes rather than inspiring learning or mutual respect.

A variant of living in diversity is the practice of cosmopolitanism as "an outlook" that resists binary categorizing and upholds the "the mélange principle" and process of hybridization (Beck 2006, 7). Instead of group traits per multiculturalism, the unit of focus is on the individual's values with an emphasis on empathy, reciprocity, and "obligations to strangers" (Appiah 2006, 97, 153). It is a commitment to pluralism of values "worth living by" and to fallibilism that recognizes "our knowledge is imperfect, provisional, subject to revision in the face of new evidence" through dialoguing and reasoning as a political community (Appiah 2006, 144). Cosmopolitanism has remained largely a normative conviction of how personal life should be lived in diversity rather than a policy measure, as in the case of multiculturalism or assimilation. Its emphasis on the individual rather than the group makes cosmopolitanism a valuable perspective on cultural diversity that avoids the narrow and essentializing characteristic of a groupist view.

Instead of multiculturalism or cosmopolitanism as means to improve intergroup relations in diversity, Amin (2002, 967) advocated for a paradigm and policy shift to a process of "urban interculturalism" that emphasized active, frequent, and local cultural dialogue between groups through common activities in publicly accessible urban places like community centres and gardens, sports clubs, and college classrooms. He explained,

> My emphasis, in contrast, falls on everyday lived experiences and local negotiations of difference, on microcultures of place through which abstract rights and obligations, together with local structures and resources, meaningfully interact with distinctive individual and interpersonal experience. This focus on the microcultures of places is not meant to privilege bottom-up or local influences over top-down or general influences, because both sets make up the grain of places. It is intended to privilege everyday enactments as the central site of identity and attitude formation.

Thus, in contrast to the abstract discursive space of national politics about immigration and multiculturalism, and intellectual debates about cosmopolitanism, urban interculturalism advocates for contact zones[16] in locales such as parks, libraries, and cafes to become strategic spaces to practice living with differences. These spaces are the ordinary realms where complete strangers and familiar strangers who have routinely observed each other from a distance without direct verbal contact per Milgram (1972) cross paths regularly. These spaces are also where "strangers become neighbours" through repeated social interactions, shared concerns, and crises (Sandercock 2000).

Intercultural living hence departs from multicultural living in its intentionality for direct engagement through contact between individuals and between groups to relate across differences with the purpose of learning by interacting.[17] When people of different cultures live or work together without the engagement of interculturalism, misunderstandings leading to conflicts could inadvertently arise because of the culturally embedded and limited worldview of each culture.[18]

Overall, interculturalism is "less groupist and culture-bound" (Meer and Modood 2012, 185) and a mode of practice that seeks "integration, interactions and promotion of a shared culture with respect for rights and diversity" (Bouchard 2015, 32).[19] However, major critiques of interculturalism are about its lack of capacity to address the power inequalities between groups, its lack of intellectual rigour, and its perception as a political myth (Meer and Modood 2012). In response to these critiques, I share Gerard Bouchard's (2015, 44) optimism about interculturalism as a progressive way forward to manage tensions arising from diversity, as he cautioned against throwing the baby out with the bathwater:

> It goes without saying that the effectiveness of intercultural dialogue is limited by power relationships, practices of discrimination, exclusionary measures and social inequalities. This is an important constraint, and it is why this dimension must be included in interculturalism, coupled with a call for social change.

Our social intercourse will always be laced with power relations that are unequal in different dimensions. To address this inequity, there is a need to establish productive means of communication that can move towards reconciliation. For this reason, I think that interculturalism offers a way to construct paths of connection between differences. Intercultural contact affirms and enables a culture of exchange and reciprocity between individuals and groups. This kind of contact proposes to create opportunities for "equal status contact" that can reduce prejudice (Allport [1954] 1979) and other forms of social injustice. Intercultural contact does not assume that social and cultural groups are static. Instead, precisely of their fuzzy boundaries and the embedded tensions arising from indeterminacy, productive interpersonal interaction via intercultural learning is put forth as a hypothesis in this book that holds the promise of developing collective life in diversity. The working definition of intercultural learning that this book uses refers to "engaging in mutual learning and adaptation between different cultures and ethnicities." The discussion of interculturalism as lived, practiced, and conceived among residents in the diverse locales will be presented in chapter 6. In chapter 7, I will draw together the findings about tensions and belonging in diversity, as well as the opportunities in the everyday public realm for interculturalism, to discuss the planning and design considerations for creating public environments in diverse locales that are productive for collective public life.

The Trouble with Diversity

Diversity as we know it in the media connotes promises of economic advantage and cosmopolitan aspirations, alongside the perils of differences that threaten to fragment social and cultural groups further. Sociocultural diversity appears to be an Achilles heel for the crafting of collective futures, particularly in contexts where the diversification process may produce multiple identities and belongings that are then manifested geographically in voluntary and involuntary residential segregation. We see the emergence of suburbs that are sorted by ethnicity, e.g., "ethnoburbs" (Li 2009), and neighbourhood choices shaped by durable structural racial biases and hierarchy (Krysan and Crowder 2017).[20] This process may also interact with existing poverty and income inequalities to form compounded social and spatial exclusions or what Douglas Massey (1996) calls "hypersegregation."

It should be observed that each individual's response to sociocultural diversity can vary. Differences, whether of sociocultural origins or socio-economic levels as perceived through the way people behave, dress, look, think, and speak, can trigger a sense of unfamiliarity that in turn generates anxiety, mild discomfort, or a sense of adventure. Diversity is not everyone's cup of tea. Ordinary contact zones may function as double-edged spaces, capable of growing antagonistic tensions between groups as well as dialogical tensions that catalyse intercultural learning and understanding.[21] The volatility of interaction in public places of intermingling and gathering in cities makes these spaces serve as critical infrastructure in the formation of constructive intergroup relations.

Los Angeles is no stranger to the kinds of insidious friction and explosive conflicts that differences can incite if they are not managed. Images of the 1992 civil unrest linger in the minds of those who lived through it. The riots spread from South Central towards midtown Los Angeles where many Koreans owned businesses and where today's Koreatown stands. Looting, fires, violence, and precariousness filled the air as the city raged against the uncountable incidents of interracial misunderstandings, socioeconomic inequity, and criminal injustice experienced in the city prior to the police's brutal arrest of Rodney King, a Black American.[22] We call to mind Rodney King's desperate address to the public as the riot continued raging in 1992 with thirty-seven dead, over 1,300 injured, 4,000 arrested and damages of over $200 million at the time of his televised appearance. It serves as a poignant reminder that coexisting in diversity is not a pleasant stroll in a park but rather a negotiated commons that requires constant vigilance:

I just want to say can we all get along? Can we, can we get along? Can we stop making it horrible for the older people and the kids? And I mean we've got enough smog here in Los Angeles. That alone and to deal with setting these fires and things and it is just not right. It's not going to change anything. We'll get our justice. They've won the battle, but they have not won the war. We'll have our day in court, and that's all we want. I am neutral. I love people of colour. I am not like that which they make me out to be. We got to quit. We've got to quit. After all, I can understand the upset first two hours after the verdict. But to go on, to keep going on like this – to see the security guard shot on the ground, it is just not right. It's just not right. Because those people will never go home to their families again. Please, we can get along here. We all can get along. I mean we are all stuck here for a while, you know. Let's try to work it out. Let's try and work it out.[23]

2 Comparing Spaces of Globalization and Diversity

A globalizing space is one that is located in a tension of difference. It is a space where we encounter groups with different social orientations, immigrants with varying statuses of residence legality, and feelings of transience and rootedness. In this space, we experience ongoing processes of arrival, departure, transfer, and resettlement. We engage with technologies that compress time and distance so that those who are near become distant and those who are physically far apart are instantaneously brought into emotional and mental proximity. Globalization has reshaped our sense of place and of time. Geographical distance has lost its salience while social distance has gained new meaning as an increased momentum of voluntary and involuntary movements of people across national and cultural borders has redefined formation of social relations in our cities, towns, and villages. In gateways of immigration in particular, social ties are destabilized, and local social space is remoulded to take on what anthropologist Arjun Appadurai (1990) coined as ethnoscapes, i.e., shifting landscapes of group identity.

Neighbourhood and Community in Tension

The notion of "my 'hood" (slang for the "neighbourhood" that emerged to describe low-income residential areas) as an assertion of belonging to a locale built on familiarity of everyday life experience has been challenged as the globalizing neighbourhood is a multi-scalar and multicultural space of habitation – at once local, international, and transnational. Familiarity is disrupted by new and different faces, colours, sensibilities, sounds, smells, beliefs, and mindsets as the local social space transforms into a transnational relational web with new and continuing flows of strangers from distant lands. New boundaries overlay old borders and new claims to place are asserted as the globalizing flow of capital, labour, and information reorders affiliations. One asks, "Whose 'hood is it?"[1] Urban territoriality begins to matter even more locally, even as geographical distance between places loses meaning (Gupta and Ferguson 1997).

The concept of the neighbourhood according to Banerjee and Baer (1984, 17) has been used as a method of "structuring, ordering, and presenting the urban society" since antiquity. Many ancient cities in Egypt, China, Greece, and those during the Roman times show evidence of the use of this concept to articulate the unit where "a group of people share a place." In 1939, American Clarence Perry developed the *neighbourhood unit* as a template for new planning and design guidelines to create an ideal and functional neighbourhood that was conducive for supporting family and local community life in rapidly urbanizing America. Some of these guidelines included the quarter-mile (about four hundred metres) walking distance for children from home to school, street layouts that enable pedestrian connection to major social amenities with minimal disruption from vehicular traffic, and the provision of adequate community spaces. Together with the social reformist ideals of Jane Addams, Robert Park, and John Dewey, who embraced the values of humanistic Gemeinschaft,[2] the concept of the neighbourhood as a planning intervention to renew local community life in cities has become widely accepted worldwide (Banerjee and Baer 1984). Over the decades, the neighbourhood has also recognizably become a common unit used in urban studies research and urban policy formulation, such as that of residential segregation and social capital in the city.

However, this entanglement of the neighbourhood as a meaningful unit of local community life, termed by Blokland (2003) as the "Siamese twins of neighbourhood and community" has created much unease among scholars.[3] The physical and social coherence suggested by the "neighbourhood" and "community" concept tends to obscure the heterogeneity found within. Herbert Gans's (1962) *The Urban Villagers* is a case in point. Through an ethnographical research of Boston's West End before its demolition as part of urban renewal in the late 1950s, Gans found that the West End was socio-economically and socioculturally diverse, as opposed to the portrait painted of it as a monolithic, poor, working-class neighbourhood. In addition, as global immigration increases demographic diversity of neighbourhoods, and technology enables us to find community in distant places, the "Siamese connection" of neighbourhood as our community draws even greater critique. At the conclusion of her study of Dutch neighbourhoods, Blokland (2003, 207) wrote the following: "This is how the neighbourhood exists ... The neighbourhood is not, never was and can never be a community. Instead, it serves practical and symbolic purposes as a means to form and perpetuate many different communities."

Nevertheless, research on neighbourhood choice has also illustrated that individuals choose a neighbourhood where they feel that they can be part of the community of residents. The notion of seeking community in a neighbourhood where an individual can feel protected, welcomed, and accepted is perceivably accentuated in a diverse and fast-paced urban life that creates a counter desire in individuals for predictability. A sense of community provided by co-ethnics from the same class or income level, who speak your language, understand your way of life intuitively, and share common aspirations can be highly desirable to some. In Aujean Lee's (2019) study of the neighbourhood choice of middle-class Latino and Asian homeowners in Los Angeles, she found that they seek out co-ethnics as neighbours for protection against discrimination, in addition to seeking out high value amenities such as good schools and safety. Similarly, in London, Watt (2009) found that the middle-class White homeowners in a suburb of East London actively distinguished themselves from their surrounding poorer neighbours by mentally carving out their housing development as separate and different. Thus, while the physical neighbourhood should never be assumed to be synonymous with the notion of a community that is socially coherent because of its shared norms and values, empirical evidence has shown that communities of ethnic, racial, class, and income coherence can be found in some neighbourhoods because of voluntary as well as involuntary residential segregation.

A globalizing neighbourhood (whether demographically homogeneous or heterogeneous) is thus a dynamic social space subject to a continuous process of identity and status differentiation by its residents, as well as between the residents and those on the outside. As philosopher and sociologist Henri Lefebvre ([1974] 1991) explained, a social space is composed of three key components, namely the practices (perceived space), experiences (lived space), and mental conceptions (conceived space) of its inhabitants and users. These three components interact continuously to produce social space, i.e., the neighbourhood. We can thus reason that a neighbourhood undergoing changes from global immigration is a complex social space that requires continual negotiation because the practices, experiences, and conceptions of its inhabitants are expected to be structured by a diversity of values, priorities, and needs.

Approaching the Three Locales

I approach the three locales as neighbourhood spaces produced by the lifeworlds of individuals in interaction with the processes of globalization, local history, and municipal planning. The neighbourhood space offers a location to understand how multi-ethnic and multinational diversity is lived, practiced, and imagined in the context of varying socio-economic conditions. Thus, the neighbourhood space is first a field site to observe, engage, and interpret how coexistence may be differently negotiated in globalizing urban settings, and is second a laboratory through which common patterns of diversity may be discerned as they manifest in interaction with other urban processes.

In line with Thomas F. Gieryn's (2006) compelling discussion in the article "City as Truth Spots: Laboratories and Field-Sites in Urban Studies," I wish to explore the neighbourhood space in this book like a mini "truth spot" to uncover and gain insights into the processes of contemporary societal transformation. However, I want to emphasize that my claims drawn from the study of only three locales in a large American metropolis cannot and do not seek generalizability. Instead, this book recognizes the complexity that accompanies diversification and the multiplicity that is common at the metropolitan scale. To this end, the interpretation of findings in the three locales would be better regarded as vignettes or metaphors of lived diversity, whereby the reader may be challenged, upon each presentation, to draw their own truths.

I have used multiple sources that informed the selection of the three locales. These sources include anecdotal evidence, personal knowledge and observation, US census data, local news, and interviews with planners in different cities of Los Angeles County. The main criteria that I used to narrow down the choice of the locales were recorded evidence of social tensions between different social and cultural groups, the lack of a majority-minority split, and the presence of multiple groups within each locale.

Central Long Beach was the first locale of the three to be identified. I found out through another research project that I was working on about Southeast Asian ethnoscapes in Los Angeles[4] that Central Long Beach had experienced different waves of immigration and there was ongoing contestation between the different ethnicities and nationalities in residence arising from the proposal for a "Cambodia Town." *San Marino* was the second locale to be selected for the study. It was chosen because of the personal conversations that I had with its former and current residents, in which the topic of tension arising between the Chinese (mostly wealthy immigrants from Taiwan, Hong Kong, and China) and White residents over differing value systems was brought up. The tensions between the diverse residents in San Marino has also been an area of sporadic journalistic interest since the 1980s.[5] Further, I thought San Marino being a high-income enclave would provide an interesting contrast of socio-economic environment to Central Long Beach, where median household incomes are much lower than the median household income level of Los Angeles County.[6] Thus, the contrast between the two helps set the research on a path of comparative study of locales with different income levels. Besides, there was relatively less research about an affluent yet diverse neighbourhood like San Marino.

The choice of the third locale, *Mid-Wilshire*, was shaped by the characteristics of the first two and the decision to establish some comparative axes that would connect the three locales. The decision to include a comparative socio-economic (income) axis made necessary a search for a mixed-income locale that was also multi-ethnic and multinational. In addition, as Central Long Beach and San Marino have Asian residents (albeit of different ethnicity and nationality), the decision was to include a third locale with an Asian resident population. Further, Mid-Wilshire as an ethnoscape or collage of ethnoscapes par excellence had been experiencing tensions arising from the contestation over spatial claims by different groups, such as the Koreans and Bangladeshis, as well as by the area's wealthy and powerful residents, who were resisting the urban changes as the locale diversified socioculturally and socio-economically.

As socio-spatial units, these locales are social spaces with tacit boundaries and territories that are expressed socially, spatially, and symbolically, with each dimension interacting with the other to reify their salience in the formation of social life in the neighbourhoods. Territoriality, according to sociologist Gerald Suttles (1972, 188), has very practical purposes and meaning because it functions "to preserve people from the prospects of insult and injury while introducing accountability into interpersonal negotiations." Hence, territoriality connotes boundary-making in that when boundaries are defined, tension is stoked because with each inch of inclusion, there is an equal inch of exclusion.

As neighbourhoods, these locales are semi-public realms or secondary territory that straddle the intimate realm of the home and the public realm of the city. They are "a blend of public accessibility and private use" that is vulnerable to misreading with regards to who and what is permissible on the ground, according to social psychologists Irwin Altman and Martin Chemers (1980, 133). Hence, in a semi-public realm where familiar strangers[7] and complete strangers intermingle, one's territory can easily be temporarily usurped by an unexpected visitor. The territorial ambiguity is perceivably exacerbated in neighbourhoods that are demographically diverse with multiple groups of different interests. Multi-ethnic and multinational locales are characterized by "grids of difference" where social and physical boundaries intersect, overlap, and contest for prominence and where identities are intertwined with place to produce territories that are both material and abstract (Pratt 1998). In fact, as Fredrik Barth incisively pointed out in *Ethnic Groups and Boundaries: The Social Organization of Culture Difference* ([1969] 1998), boundary construction is a continuous process of differentiation between the inclusion and exclusion of groups. This dynamic process unhinges the study of ethnic groups as fixed notions, making its study more socially and politically important.

Boundary formation and negotiation is dynamic, imaginably more so in spaces of globalization and diversity where the process of boundary formation itself fleshes out how social and cultural differences are negotiated in space and through the medium of space. The recognition of this dynamism presents a major challenge for the research design of this study: should I define actual geographical limits of the three locales so that data can be systematically collected for comparative analysis? If so, how do I go about it without undermining the dynamic process of boundary formation?

As the major objective of the research is to understand how space in the neighbourhood is shared in diversity and to investigate the possibilities for intercultural learning in public places, I sought out geographical areas within the three locales that had nearby accessibility to civic places of gathering. In this case, I selected those with at least a public park and a public library. In addition, the study undertook several measures to avoid fixing a neighbourhood boundary from a top-down researcher perspective. I conducted preliminary interviews with community organizers and local residents to understand how space was organized on the ground. I made observations of the built environment character and street activities through multiple site visits and consulted a variety of secondary sources including newspaper articles, local government documents, and zip code boundaries before deriving a provisional study boundary. The delineation of the study area boundaries was made particularly difficult because census tract boundaries did not line up with zip codes and available secondary documents about Central Long Beach,[8] San Marino,[9] and Mid-Wilshire.[10]

Delineating study boundaries clearly has its trade-offs, but it also enables a spatial entry point to be established for the research. To minimize a fixed set of predetermined boundaries, participants were asked to map their set of boundaries (if present) of what they geographically define as their neighbourhood on Google Maps during the cognitive mapping interview. In so doing, the intention was to let the participants define their personal space of reference and relevance to the concept of the neighbourhood. These sets of boundaries, when collectively analysed, could also provide research insight on how individuals living and working in a globalizing and diverse locale practice and imagine their space – whether there are fuzzy or clear boundaries, and/or common borders and territories.

Cognitive Mapping of Diversity

The original method employed by Lynch required participants to produce a sketch map of the city on a blank sheet of paper, in addition to detailed descriptions of routines used by the public to travel through the city. For this study, instead of Lynch's ([1960] 1998) method,[11] I used street maps from Google Maps as a base on which cognitive mapping was undertaken. The decision to use street maps was made to minimize the reliance on the participants' ability to sketch, and to enable an easier way to compare the maps between individuals and across the three locales. However, admittedly, there are also trade-offs in using a street map instead of a blank sheet: compromises on the opportunity to compare the contents of the different images, which could provide information about an participant's interpretation of how they experience their lived space, and on the opportunity to compare different expressions in which the area is imagined through the way and sequence a neighbourhood is drawn.

Overall, the cognitive mapping interview process was very effective in orienting the participants to talk about their neighbourhood as a physical and social space. In addition, the mapping exercise functioned well as an icebreaker to ease the participants into a conversation with me, a stranger. Some of the participants told me after the interview that the mapping process was very engaging and enjoyable.

To operationalize this form of "Lynch-inspired" cognitive mapping, two Google Maps – one at a "local" scale of 1:12,500 that included the area of interest and the immediate surrounding street blocks, and another at a "regional" scale of 1:25,000 that included surrounding neighbourhoods – were given to each participant at the start of the interview. This was done to mitigate imposing a set of rigid geographical limits of a neighbourhood on the participants by the researcher. In retrospect, this decision was a necessary and good one as participants across the three locales demonstrated a discernible pattern in their choice of the maps that have helped to further validate and strengthen the findings on the size of the social space in each locale, and how it is perceived, conceived, and lived under different conditions of diversity. Most of the residents across age groups in San Marino selected to map their social space at the local scale, while regular visitors to San Marino from the surrounding cities used the regional map to locate their neighbourhoods and explained their impressions of San Marino. In Central Long Beach, almost all the participants used the local scale map, except for a handful who made use of the regional map as well. In these cases when both maps are used, the participants were usually young White and Black residents who spoke English well or community organizers. In the case of Mid-Wilshire, about half of the participants used the regional map to discuss the social space of the locale for them, with some using both maps to offer more details. The characteristics of residents who chose the regional map were those in their twenties to forties, and who also owned automobiles. I will discuss more on this matter in chapters 4 and 5 as I present the findings of the cognitive maps and interviews on tensions, boundary-making, and territorial formation in diversity.

After the participants had chosen the map, they were asked to locate where they lived with a star sticker on whichever map that made most sense to them. Participants were then given coloured markers and asked to draw the boundary of their "neighbourhood" on the map and mark the locations of the areas they felt were unsafe, as well as the locations or the areas of concentrations of social and cultural groups that they had personally experienced or knew the existence of. Please see appendix 3 for the interview questions that were used to guide the cognitive mapping process. See figure 2.1 for samples of maps at two different scales by residents from the three locales.

While most of the participants were able to mark the geographical extent of the area they lived in, some required more explanation of what a "boundary" means. I was aware to minimize my influence as much as possible, limiting my description to "you can circle, highlight the streets, or box it." On the same set of maps, I asked participants to mark out their routine destinations and locations of intergroup encounters in these spaces. These individual maps were then transferred digitally to facilitate analysis of collective patterns. Initially, I analysed and compared these maps using the filters of ethnicity, nativity status (i.e., native or immigrant), and length of residency. During the process, I realized that my sample size of each ethnic group represented in the locales was too small and inconclusive. Of the three filters, the length of residency yielded more conclusive findings, such as between long-term and short-term residents in Central Long Beach, or

Figure 2.1. **San Marino:** (1:12,500 map) A first-generation Asian American resident in her fifties who lived in the western part of San Marino (star sticker) marked out her neighbourhood boundary in a large polygon and a smaller polygon to show the concentration of the wealthy in San Marino.

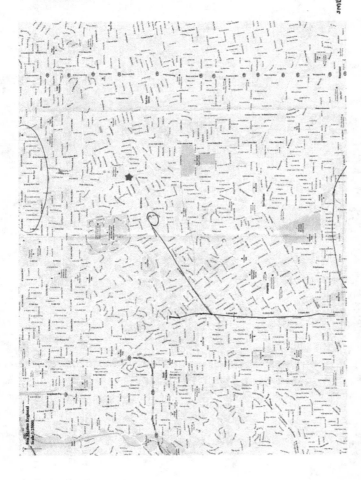

Figure 2.1 (*continued*). **San Marino:** (1:25,000 map) A young Chinese international student in his twenties who regularly visited San Marino for its library (star sticker) traced out the route that he drove from Monterey Park to San Marino each day. The two smaller polygons indicate the neighbourhoods of Pasadena to the north and Monterey Park to the south of San Marino, which he explained are distinctively separate and different from San Marino. In both maps, the participants indicated their regular visits to their friends' homes in San Marino with a small circle.

Figure 2.1 (*continued*). **Central Long Beach:** (1:12,500 map) A young early twenties Latino/Hispanic American marked out in a rectangle the area that he considers the larger neighbourhood of Central Long Beach for him, within which he drew a smaller polygon of his neighbourhood. His rented family home was located with a star sticker. Outside his home, he marked a dangerous street to the right and to the left, and homes of friends that he visited regularly. Other small polygons south on the map indicated places he visited regularly.

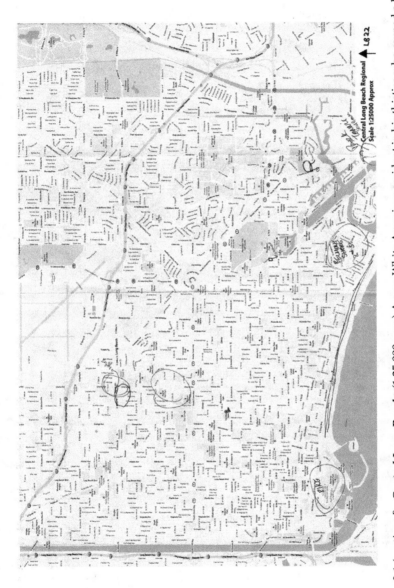

Figure 2.1 (*continued*). **Central Long Beach:** (1:25,000 map) A new White American resident in his thirties who owned a home in the area mapped his wider social space outside his immediate neighbourhood of Central Long Beach. The polygons and lines show the places he frequented on a regular basis in the City of Long Beach.

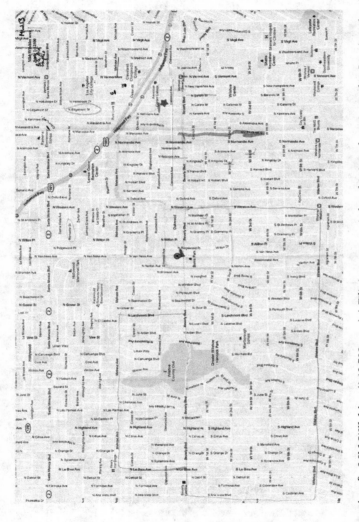

Figure 2.1 (*continued*). **Mid-Wilshire:** (1:12,500 map) A Filipino woman in her thirties who migrated to Los Angeles about four years earlier located her apartment with a star sticker. The irregular polygon around her apartment was her neighbourhood boundary. The two large rectangles to the left show the neighbourhood of Wilshire Library where we had the interview. Note that there is a small overlap between the two sets of boundaries at the library (marked as A). She considered the thick line south of the apartment to be a dangerous street. She also circled street names where she perceived strong ethnic concentration of Latinos.

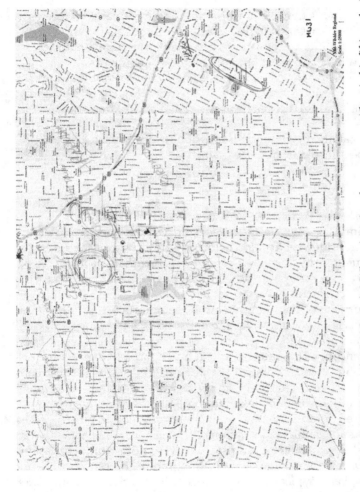

MU31
Mid-Wilshire Regional
Scale 1:25000

Figure 2.1 (*continued*). **Mid-Wilshire:** (1:25,000 map) A White American homeowner resident in his fifties who had lived in the area for about two decades mapped his social space in detail. The large irregular polygon was his neighbourhood boundary, while smaller polygons to the west and north of his neighbourhood were ethnic and lifestyle concentrations, for example, Russians, gays, and actors in West Hollywood, and Hassidic Jews to the south. He also circled areas of danger north of his neighbourhood as well as to the east, in Downtown LA.

between residents and regular visitors in San Marino to the locales. In the case of Mid-Wilshire, however, the location rather than the length of residency mattered most. The analytical maps are found in chapters 3 through 5. Overall, the analysis was guided by the following questions across the three locales:

1. How do residents living in multi-ethnic and multinational locales conceive the boundaries of their neighbourhoods?
2. How is the neighbourhood space occupied? Are there territories, and if so, where and what kind of territories are they?
3. How do the routines of residents relate to the set of boundaries and territories outlined by them in the neighbourhood? Do these boundaries and territories limit the range and pattern of routines?
4. Finally, what are the common issues about social space formation that can be inferred from these maps?

Reflections on Fieldwork

The fieldwork was completed over two phases.[12] In May 2011, I conducted preliminary interviews in each of the three locales. After the preliminary interviews, I decided to include a short survey at the end of each subsequent semi-structured interview – one that specifically asks for opinions about the types and qualities of space that are conducive to social interaction and intercultural learning.

For the purpose of comparative study, semi-structured rather than in-depth interviews about intergroup encounters, experiences of social relations in the area, and intercultural opportunities were used. Even so, quite a handful of the interviews felt much more like in-depth conversations about the participants' life stories and decisions. These conversations lasted for three to five hours, much longer than the average one-and-a-half hours usually taken to complete the cognitive mapping, semi-structured interviews, and a short survey about public places and basic demographic information. Ethnographic interviews were also undertaken whenever circumstances were not suitable for a longer semi-structured interview. These ethnographic interviews were not voice-recorded and typically lasted fifteen minutes. See appendix 3 for the list of semi-structured questions and the short survey.

Between August 2011 and February 2012, I interviewed residents, visitors, business owners, community organizers and municipal officers. I participated in neighbourhood events (e.g., fundraisers, a Cambodian Arts festival, a Martin Luther King parade, and farmers' markets) and attended neighbourhood meetings. In addition, I also interviewed two experts on human relations in Los Angeles during the course of the fieldwork. I began my interviewing in San Marino, followed by Central Long Beach and then Mid-Wilshire. The fieldwork in each locale took about two-and-a-half months to complete, with overlapping weeks as I transitioned from one locale to another. I conducted 100 interviews lasting at least an hour each and another forty ethnographic-style, shorter interviews during events, meetings, walking, and visiting the neighbourhood shops and public spaces.[13] Overall, shorter and less structured ethnographic-style interviewing without recording was used most frequently in Central Long Beach. Possibly due to genocide trauma and their experiences as refugees, many Cambodian participants did not feel comfortable partaking in a "formal-looking" study with recording and note-taking.[14] In total, 140 interviews were held, and seventy-eight of the participants participated in cognitive mapping. In addition, sixty-eight surveys were filled out.[15] Table 2.1 shows the total number of interviews in each locale.

Conducting fieldwork in three very different sites over eight months using commuter fieldwork necessitated continual sociocultural border crossings on my part even though these locales were all in the Los Angeles metropolitan area. Each locale presented a unique set of entry dynamics including the requirement to navigate different municipalities, organizations, social networks, social behaviour, and norms of approaching strangers in each locale. What seemed like a logical sequence of questions, received well in one study area, was rather awkwardly received in the other two. Thus, the preliminary interviews were extremely helpful to flesh out these differences to inform the modus operandi that I should employ in each locale, and to prepare a set of questions that could be broadly used across

Table 2.1. The composition and number of interviews in each locale

Study areas	Semi-structured interviews	Ethnographic interviews	Total	Surveys completed	Cognitive maps*
San Marino	30	14	44	(21)	(25)
Central Long Beach	34	19	53	(23)	(26)
Mid-Wilshire	34	7	41	(24)	(27)
Interviews with human relations experts	2	0	2	(0)	(0)
Total	100	40	140	(68)	(78)

*This is an approximate count because some participants were only able to partially complete the cognitive mapping interview.

the different locales. In addition, a significant amount of time was spent scheduling interviews and "sowing the seeds" in each locale for future interviews so that the fieldwork could proceed without too much time-lapse between sites.

During the period of the fieldwork, I commuted to the locale two to three times per week to interview residents and to observe the activities in the public places. Some might argue that such fieldwork lacked ethnographic authenticity because I was a perpetual outsider, merely commuting to a research site rather than immersing in it, particularly when diverse locales are inherently socially complex with multiple groups and alliances. However, a lack of full immersion has also afforded a clarity and some level of neutrality in these locales where intergroup tensions are rife.

Recruiting Participants

My pool of participants includes residents and regular visitors of the neighbourhoods, business owners, community organizers who work locally in neighbourhood civic organizations (religious, social services) and city services (park, library, police), and the municipal decision makers (e.g., planners, district representative) in the city hall. Among these categories of participants, the residents and regular visitors were the hardest to recruit as no interview appointments could be set up ahead of time. So, I sought out the neighbourhood library as my base and entry point into the three locales. With the support of the librarians, I was able to conduct the interviews in the library. In Central Long Beach where there were few public areas that felt safe, quiet, and diverse, the library was a good location for the interviews. In Mid-Wilshire, the library was a location that attracted an ethnically diverse crowd living in the locale and thus, it was a fruitful recruitment site. In San Marino, very few residents used the library, but it proved to be a helpful location to recruit participants who are regular visitors to San Marino from the surrounding cities. Most of the resident participants in San Marino were recruited from interpersonal social networks and referrals from serendipitous meetings with people who lived there or who had friends living there.

To recruit participants, I had to personally approach strangers in the libraries and in parks to ask if they would have forty-five minutes to one hour to assist me in my research study. I had a fair share of rough rejections and polite ones. Admittedly, this mode of recruiting could be critiqued for its lack of representativeness because of the self-selection process between the interviewer and the participant. To minimize narrow selection biases and maximize diversity of viewpoints, I sought to recruit participants through various sources and from different venues including public spaces (libraries, parks, community centres) and civic organizations (religious and social services organizations), as well as from a variety of interpersonal social networks in each of the three neighbourhoods. I also used the snowballing recruitment technique to ask participants to refer me to their friends or people whom they thought would be agreeable to be interviewed. However, the potential of snowballing as a process to gain new contacts through these interviews at public spaces was extremely low, although not completely absent.

Researcher Positioning: Insider, Outsider, or In-between?

Undertaking this research about ethnicity and spatial behaviour as a foreign student from Asia, in the immigrant gateway metropolitan Los Angeles, presented practical opportunities and constraints as well as insights and blind spots. As a fourth generation ethnic Chinese who was born and raised in Singapore, I saw myself as an overseas Chinese with more cultural connections to Southeast Asia than to China. For this reason, when asked, I would avoid introducing myself as "Chinese," preferring to describe my identity as an Asian student from Singapore studying at the University of Southern California. Due to my physical features, I was frequently mistaken as Thai, Filipino, Korean, Latino, or Guamanian. It was a double-edged sword, and its benefits and harms were hard to discern and isolate. I was welcomed into some places and conversations like an insider, while I was also thanked by several participants for being an outsider who did not belong to any of the groups represented, which allowed them an opportunity to process their thoughts about inter-ethnic relations objectively and openly. I was also rudely waved away like a telemarketer or a saleswoman by others, usually by Asian men who did not want to be interviewed by an Asian woman speaking in English.[16] One could only guess at all the possible reasons for social acceptance and rejection arising from the intersectionality of identities! I reflected upon my privileges, limitations, and identity negotiation in each locale below:

In Central Long Beach, looking Southeast Asian allowed me easier access to the Cambodians and Vietnamese living and working in the area. Of note was the warm welcome that I received by Cambodian seniors in the neighbourhood recreation centre (many of them were overseas Chinese Cambodians), who greeted me with enthusiasm and asked if I spoke Mandarin and the southern Chinese dialects Teochew and Cantonese. As I could understand the three and speak Mandarin and Cantonese, the Cambodian Chinese seniors were delighted. Over lunch, I was able to ethnographically interview these elderly Cambodians using a mix of Mandarin, Cantonese, and English. They afforded me insights into the experiences of being Cambodian refugees and their feelings of ambivalence regarding their identity as Cambodians who felt a greater connection to China than to the United States and Cambodia. As I pondered over this and contrasted their feelings of being an overseas Chinese in Southeast Asia with my own, I realized that their strong ethnic ties to China (as compared to my weak ties) could be a result of the minority status that Chinese experience in Cambodia, among other reasons.

On the other hand, I had the most social angst conducting fieldwork in Central Long Beach. There was no "comfort zone" of a cafe space or a nice park bench to hang out, at least not until a new Subway restaurant opened up on the northern boundary to provide me a spot of respite to reflect and recharge in the field. I felt out of place even though I look Southeast Asian or Hispanic, two groups which together form the majority in the area. It did not help that I also stood out during the times when I was taking pictures and disrupting the banal humdrum of daily life. The ethnographic process of "hanging out" and walking the streets was also socially awkward in a neighbourhood that limited lingering in public space where unpredictable gang violence could happen in the streets. Due to all these reasons, the library provided a bona fide interview space where my identity as a researcher was acknowledged, my presence was approved by the library's authority, and where I could be relaxed to approach potential participants.

Overall, my inability to speak Spanish limited my access to interview Latinos, many of whom are immigrants from Mexico, Guatemala, and El Salvador. My lack of Khmer-speaking ability also limited the depth of the interviews with Cambodians who were uncomfortable speaking English. However, my experience working with internationals whose first language was not English helped me to adapt the questions for these participants. Among those whom I interviewed, I found good rapport with Black American participants. Unencumbered by language, the interviews enabled conversations that explored deeper nuances of race and intergroup relations. My status as a minority, an Asian, and a woman could have also provided a "familiar" and "neutral" position that helped the Black American participants to relate to me more easily. A few of them thanked me for helping them to process, think about, and articulate their feelings openly about race.

In San Marino, my identity as an overseas Chinese and my ability to speak Chinese and English were extremely helpful in opening doors to new Chinese immigrants from China, second-generation Asian Americans, and first-generation immigrants from Taiwan and Malaysia. Perceivably, these "advantageous" personal features could have disadvantaged me when interviewing White Americans about their inter-ethnic relations with the Chinese. However, I felt that my non-resident status, Singapore nationality, together with my University of Southern California student researcher identity, were perhaps more prominent as characteristics of differentiation than ethnicity per se. The White American participants in San Marino were candid about their inter-ethnic experiences with the Chinese and if there was any initial discomfort, I felt it dissipate as the interview progressed.

In Mid-Wilshire, my Asian ethnicity was less of a disadvantage or advantage as compared to my position as a female researcher, although I could not discount its influence in opening doors to Korean and Filipino female participants who might be predisposed to feel more comfortable talking to me. As a female researcher, I was able to access certain places such as playgrounds, "hang out" in parks to look for participants without being questioned, and could easily approach mothers and nannies without creating unease. Similarly, to prevent my Asian identity from limiting openness in the discussion of social relations among Asians in the area, I would address this ambivalence upfront by informing the participants that I was from Singapore, i.e., not Korea or the Philippines. Like in Central Long Beach, my lack of ability to speak Spanish limited my access to Spanish-speaking nannies and mothers in Mid-Wilshire, many of whom did not want to chat or be interviewed.

Admittedly, the use of English in multilingual sites to conduct interviews limited the data collection to English speakers and thus, the sample could not be as representative of the composition and diversity of the actual population living in these multi-ethnic and multinational locales. However, using English as the main interview language introduced a useful control into the research to tease out the other barriers (apart from language) that made intercultural learning and engagement difficult.

I will remember the oft exhausting social awkwardness of always having to negotiate my class privilege one way or other, my accented English as a foreigner, and the need to negotiate the different types of cultural sensitivity in a diverse landscape dotted with unspoken stereotypes and divided by multiple intersecting lines of segregation. To conclude, I want to quote from Susan A. Phillips (1999, 96) at length here on her reflection on her fieldwork research about graffiti and gangs in Los Angeles as I share some of her experiences in the field:

As with most major cities, it is the nature of Los Angeles to segregate people. This segregation makes you feel comfortable on your own turf and uncomfortable on somebody else's. Because of the city's size (its famous sprawl), such zones of comfort can be enormous but still manage to exclude entire populations from their midst. I had to develop survival mechanisms for the hatred I encountered when I crossed those boundaries. I certainly felt exhilarated when I did so successfully – when I did fit in and felt welcomed and accepted, and even wanted. Ultimately, the power of those moments made it possible for me to do fieldwork in a city where divides of a few miles sometimes seemed greater than for those separating nations.

3 Scenes of Diversity in Wealth, Poverty, and Inequality

I think LA is the best example of what diversity can look like. It is not just about physical appearances, but it is about being more open to embracing the culture. You can't avoid diversity. You drive by and you see the different kinds of fruit from South America. How by adding a little brown sugar, it changes the flavour … Being different is not bad! You got to be open or else you will become isolated. Soon, it becomes more difficult for you to live.

Nancy Lau, a Chinese American in her sixties
Mid-Wilshire resident

Los Angeles is a metropolis of myriad contrasting lifeworlds that challenges settled notions of diversity. Diversity, as Nancy described above via a personal encounter, has a rich cinematic quality that engages the senses and is situated in the everyday lived experience on the streets of the city. In this chapter, I have deliberately chosen to describe the different textures of diversity in three "scenes" with the aim to bring the reader along with me on a journey to sense the unevenness and patterns of coexisting in the presence of differences. Journeying through the scenes of lived diversity conveys an element of passing time and discovery that follows closely the order which I have conducted my fieldwork, notwithstanding that "scenes" simultaneously evoke the incompleteness and freezing of time that are framed and curated through my lens.

Scene One: San Marino

On a late sunny May morning, I arrived in San Marino for the first time to begin my fieldwork. I left the noise and asphalt of the City of Los Angeles behind and entered Huntington Drive, a wide and well-paved avenue of lush green trees, generous road median and sidewalks. The quiet ambience of an established and well-heeled suburban city with green, well-watered, manicured lawns and clean streets surrounded me. San Marino stands out from its surrounding towns in Los Angeles's San Gabriel Valley as an exclusive, manicured, tree-lined, and low-density residential area.[1] This exclusivity is produced through stringent ordinances on trees and regulations on the form, façade, and function of buildings introduced by the City of San Marino to "preserve neighbourhood character and protect property values" as envisioned by its founders 100 years ago in 1913.[2]

According to Nicolaides and Zarsadiaz (2017), the aesthetic and character of San Marino were carefully crafted in the early years of 1920s by setting high minimum home prices, focusing on European-influenced residential architecture built by well-known local architects, and putting in place a land use zoning that prioritized single-family homes only with a minimal commercial presence. In addition, a Planning Commission was also formed in the 1950s to safeguard the residential guidelines laid down since its founding. Further, San Marino's racial character was purposefully maintained. Akin to 47 per cent of residential neighbourhoods in Los Angeles County in the 1930s that upheld race restrictive covenants, land in San Marino was not allowed to be sold to Jews, Blacks, or Chinese.[3] By the early 1970s, even after the federal Fair Housing Act of 1968 had outlawed racially restrictive covenants, San Marino was kept racially White as realtors steered prospective non-White buyers away and homeowners refused the sale.

Although the physical look of San Marino has been preserved thus far by its strict planning guidelines, San Marino's social space has evolved dramatically over the decades. Beginning from the mid-1970s and accelerating through the '80s and '90s, San Marino underwent significant demographic diversification. It transitioned from a predominantly White American suburban city known for its segregationist practices and being the location of the former headquarters of the ultraconservative John Birch Society from the 1960s to 1989, to a city with 53 per cent Asian residents by 2010 and 61 per cent by 2020.[4] The population increase of wealthy Chinese immigrants from Taiwan, Hong Kong, and China, and second-generation Asian Americans in San Marino over the decades led to a new ironic nickname for the city, "Chan Marino," a colloquial reference by its residents to reflect the popular Chinese last name among its new residents (Hudson 1990).

Between the 1980s and 1990s, fights in schools between children of Asian immigrants and those of White Americans were not uncommon (Reinhold 1987), perhaps mirroring and manifesting the unease resulting from the destabilized status quo in the parents' social world outside the school environment. Noah Yu, a second-generation Asian American resident who attended a public school in San Marino in the 1990s, explained to me that all the White Americans formed one social group, and the new immigrants, known as "Fresh-off-the-boat" (FOBs), formed another. The Asian Americans, America-born-Chinese, and second- or third-generation Asian Americans formed an in-between group, comprising those who spoke English and got along well with the White Americans. However, the Asian American children would deliberately shun the new immigrant children, whom they thought were rude troublemakers. These different groups practiced cafeteria segregation as students would gather in separate clusters according to these social categories. Though accepted as a common trait of school life for many, the segregation was not without friction, and fights between groups were common.

During the 1990s, two organizations were set up in San Marino to manage the crisis in souring social relations between the White American and Asian immigrant residents. The Human Relations Committee organized a program called "Dinner for Eight" that paired two Asian immigrant couples up with two White American couples, with the aim to encourage interaction and help the immigrants integrate into their new social environment. But the initiative fell apart after two years due to the lack of interest and commitment by residents, according to some of the participants who knew about this initiative or participated in it. The other initiative was the formation of the Chinese Club of San Marino. Initiated by a small group of Taiwanese residents, the club's purpose was to organize and mobilize support for the interests of the minority Taiwanese and Chinese residents (at that time) who were facing discrimination in the public schools and in their access to public resources in San Marino. During the time of this research, the Chinese Club of San Marino continued to be a powerful advocate and intermediary for its Chinese and Taiwanese members, undertaking formal and informal negotiations with the non-Chinese groups in the city. For example, the club has run Chinese language classes for city employees, facilitated city operations by translating, disseminating, and educating new Chinese residents about the building and environment ordinances in the city, and organized the annual fundraising event for the public schools in San Marino.

By 2010, San Marino was definitively described to have "lost its whiteness"[5] as the proportion of Asian residents grew to above 50 per cent of the total population. San Marino has become the first destination of wealthy Chinese professionals and businessmen immigrants, who have moved straight into purchasing and living the American Dream – ownership of a suburban single-family home with a lush and spacious backyard. As a node situated in the globalizing flows of finance, information, and peoples, its million-dollar real estates were advertised in China and Taiwan as status symbols. In fact, visits to San Marino were frequently a part of the investment tours organized for foreign Chinese businessmen who were looking for opportunities to invest in America (Ni 2011). According to Chinese real estate agent YanYan Zhang, whom *LA Times* reporter Lauren Beale interviewed, "If you go to mainland China and someone asks, 'Where do you live?' San Marino represents that you are wealthy" (Beale 2011).[6] Please refer to appendix 1.1 for the demographic information of San Marino in 2010 and 2020.

San Marino is quickly becoming an "ethnoburb" par excellence. Geographer Wei Li (2009) first identified the formation of ethnoburbs (ethnic suburbs) in the 1990s in Southern California. These ethnoburbs bucked the well-established pattern and model of American urban development conceptualized by Park and Burgess ([1925] 1967). In this model based on field studies of Chicago's urban development patterns, poor immigrants would first arrive and settle in the inner city because of the abundance of cheap rental units and social networks to gain employment. Only after these immigrants had moved up the socio-economic ladder and assimilated would they move out of the city into the suburbs. However, as Nicolaides and Zarsadiaz (2017, 351) argued, San Marino has a different place identity from the surrounding ethnoburbs of Monterey Park and Arcadia so that it is sought out as a positional good among the homeowners because it is "superior for its lush trees, good schools, upstanding kids, and the polite culture bred by high-level professionals and business people." Figure 3.1 illustrates the Asian ethnoburbs of San Gabriel Valley surrounding San Marino.

From a small exclusive American town, San Marino has become a global node where elite members of the "network society" à la Castells ([1996] 2000), such as CEOs of global companies, live side by side with Los Angeles's politically powerful, doctors, lawyers, bankers, and businessmen from some forty nations. As a "space of flows" of the mobile and footloose in this globalizing economy, San Marino is seen as a space of finance and convenience to grow the network capacity of capitalists who are too transient to invest in the place or its people.[7] San Marino is also a "space of places" which elites have invested in to make a home to raise their families. Caught in the tension between jet-setting and home-making, between exchange and use value, and between other conflicting cultural values presented in sociocultural diversity, the social space production in San Marino is highly contested.[8] As John Shaw, a municipal officer who has worked in San Marino for a decade, described during the interview, San Marino is a city where "invisible fences between groups" are built and that the uneasy quiet ruptures every so often. Figure 3.2 shows images of San Marino.

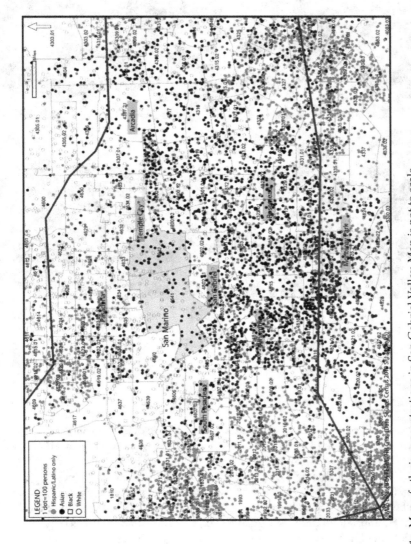

Figure 3.1.   Map of ethnic concentrations in San Gabriel Valley. Map is not to scale.

Prepared by author using US Census 2010 data SF1 Table P5.

Figure 3.2.   Photographs showing the social landscape of San Marino.
(*top left*) The tree-lined Huntington Drive formed the backdrop of a local
diner "Colonial Kitchen" that was owned and operated by an immigrant
couple from China. (*top right*) A single-family house for sale by a Chinese
realtor. (*bottom left*) Parents and nannies watching their children practice for
Little League in Lacy Park during a late weekday afternoon. (*bottom right*) The
annual Hauntington Fundraiser Breakfast organized by San Marino's only
Middle School, where Chinese breakfast staples of steam buns and scallion
pancakes were available alongside American breakfast staples of sausages and
pancakes.

Taken by author in 2012.

Socio-Spatial Differentiation: Wealth and Ethnicity

Although outwardly seen by others as a homogeneous enclave of wealthy residents, San Marino is in fact conceived as a space of social differentiation by its city officials and its residents. A municipal zoning map of lot sizes spatially differentiates the city into classes of residential zones (see figure 3.3). For example, the enclave of the "crème de la crème" residents of San Marino live in District IE with lot sizes of 60,000 square feet (about 5,500 square metres) and above and in District I with lot sizes of at least 30,000 square feet (about 2,800 square metres), according to Jennifer Meier, a resident in San Marino who was also a real estate agent. Another resident, Lydia Li, described District IE as the place where affluent "women of the hills" resided. The socio-spatial differentiation among its residents was already salient even in the 1970s. Mary Philips, who grew up in San Marino in the 1970s, informed me that smaller properties had then been referred to "South of the Drive" to indicate the smaller lot sizes south of Huntington Drive. In addition, the label "B-tract" had been used by residents to refer to the smallest and most modest properties in the far southeastern corner of San Marino (District VII).[9]

According to John Shaw, immigration was rapidly changing the status quo as far as who lived in these areas. For example, the medium-sized homes (with lot sizes in the range of 15,000, 20,000, and 30,000 square feet, or 1,400, 1,900, 2,800 square metres) around Lacy Park used to represent the "old money" of White Americans and a few Latin Americans. But it was fast becoming increasingly Asian-owned. The area adjacent to San Marino High School, where lot sizes ranged between 12,000 square feet (1,100 square metres) and 20,000 square feet (1,900 square metres), had become an area receiving the "new money" from immigrants arriving from China, Hong Kong, and Taiwan.

In an excerpt of an online video Q&A with Dr. Richard Sun, the then mayor of San Marino in 2012–13,[10] he illustrated that the social space of San Marino was divided into several social and cultural groups (*in italics added by author*) along the lines of ethnicity and nationality.[11]

SAN MARINO PATCH REPORTER: The population of San Marino is comprised of a little over 50 per cent Asians, Asian members. And of that population, the majority is Chinese. How do you think the relations are currently between the Chinese community in San Marino and the non-Chinese community?
MAYOR SUN: At the present time based on 2010 census, we have about more or less like 13,000 population in our city. In other words, out of 13,000, 53

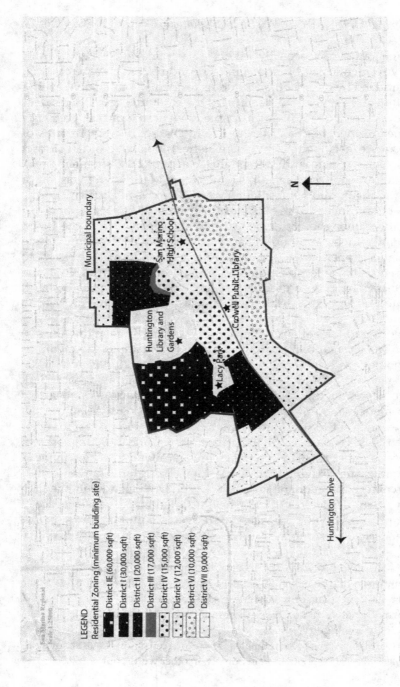

Figure 3.3. Residential zoning map of the City of San Marino overlaid on the Google Map. Map is not to scale.

Adapted by author.

per cent are *Asian Americans*. Out of the 53 per cent, 40 per cent are probably *Chinese Americans*, doesn't matter whether you come from China, or Taiwan or from other countries. At the present time, the relationship between the *Chinese community* and *non-Chinese community*, I think it just cannot be even smoother … I don't know if you have ever attended the Mid-Autumn Festival event. Basically it is a community event. The Chinese community invited the non-Chinese community to mingle together and vice versa. Like City Club's events, the Chinese community always attends the event too. We do see the communities can mingle together. And also last year I took a delegation visiting Taiwan including our city residents, including non-Chinese residents, even our city officials, chief of police, city manager and their spouses going there too. And this year I am taking another group of delegation to visit Danshui District in new Taipei City and to formalize the sister city relationship between City of San Marino and new Taipei City of Taiwan. So those things are good because we are promoting cultural exchange to have a better understanding.

SAN MARINO PATCH REPORTER: All right, is there a way you would improve the relationship in any way between the Chinese community and non-Chinese community?

MAYOR SUN: I think communication is the best way. So, whenever you open yourself, you feel like you're part of a community. You communicate not just within your own group. We encourage especially *new immigrants* to go out, to meet more people, meet your neighbours, meet people even if you may not be fluent in English but it's okay. Say hello to them. Open up your communication. Those relationships will be improved.

Likewise, among many residents in San Marino, demographic diversity was simplified as binary and mutually exclusive categories of the Chinese and the non-Chinese community at the time of research. In reality, within the "Chinese community," the Chinese participants did not view themselves as a coherent entity as suggested by the former mayor in the excerpt above. The Chinese community was stratified and purposefully differentiated. The political tension in East Asia between supporters for a democratic Taiwan and Hong Kong and those for the incumbent style of rule in China was alive as well in San Marino, as seen in the formation of the different national and subnational groups. Besides being divided by political inclinations, the groups were also divided by the period of their arrival in San Marino and their nationality. For example, the early immigrants came largely from Taiwan and Hong Kong, who had left East Asia in the 1970s to the 1990s, while the "newcomer immigrants" came mostly from China. These categories ordered the social interaction and the opportunities for it.

According to Luke McDowell, a resident and municipal officer in San Marino, another division was evident between "old time San Marino folks" who moved into the city in the 1950s (i.e., White American retirees) and the "newer arrivals" made up of Asians (i.e., younger Chinese/Taiwanese/Hongkonger and others). Naomi Su, an Asian American resident of San Marino, explained that one reason for the tension between the two groups was that they did not socialize with one another because the White American retirees did not participate in the activities of the San Marino public schools. In addition, the White American retirees and young Chinese newcomers were also culturally and generationally more distant and had less in common. Naomi who was an active member in the public school parent-teacher association (PTA), informed me that there was also a worrying trend of a widening division between the newcomer Chinese immigrant parents and the parents who are Asian American or White American.

Divergent Routines

Kept intentionally as a predominantly residential area by land use zoning and regulations, San Marino's quotidian street life is quiet and sparse, save for the commercial activities at the local Starbucks cafe, pizzeria, and the couple of small eateries near the city hall where residents can intermingle. According to the routine maps of residents, many of them regularly travel outside San Marino for their shopping and socializing needs. Most of the White American residents and second-generation, younger Asian American residents travel to Pasadena to the north of San Marino where hip cafes, Indian restaurants, Thai restaurants, French bakeries, popular departmental stores, and boutique shops are located. In contrast, the routine geographies of Chinese residents, particularly the first-generation immigrants, extend southward to Valley Boulevard, where a major thoroughfare of mainly Chinese and some Vietnamese restaurants and supermarkets are located in the San Gabriel Valley, also popularly known as part of the "China Valley" of Los Angeles. Figure 3.4 shows the collective routine geographies of White American and Asian American residents, taken from their cognitive maps.

Neighbourly interactions are rare because of long working hours, frequent travel, the desire for privacy, absentee home ownership and, simply, the different timetables of many residents who are transnational elites commuting in a global "space of flows" for work and pleasure. Furthermore, there are few possibilities for residents who do not have children attending the public schools to engage and interact with those who do not share similar demographic characteristics or background experiences. An example was Linhui Kao, a first-generation immigrant from

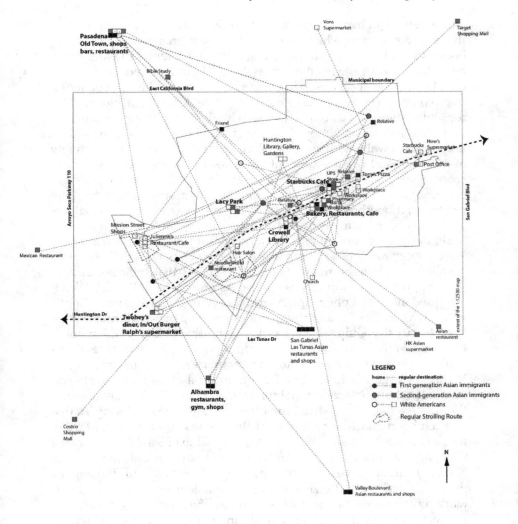

Figure 3.4. Mapping routine geographies of San Marino participants (N=14). Map is not to scale.

Prepared by author.

Taiwan who had lived in San Marino for over twenty years. For Linhui, the only occasion in which she would interact briefly with her neighbours was during her regular morning walk. She described, in Mandarin, her relations with her neighbours as "淡淡" [dan dan] meaning "bland bland," nothing special or particularly deep despite their frequent "hi-bye" contact. The Asian residents and White American residents also recounted their contact opportunities with their neighbours as typically limited to routinized "hi-bye" greetings along the sidewalks near their homes when walking their dogs, taking a stroll, or sometimes fleetingly in the front yard when getting into their respective cars on their way to work. There were nonetheless a few memorable occasions when longer conversations took place, usually about the common matters of the street or the neighbourhood. The rarity of these spontaneous conversations about collective life made these exchanges unique and memorable.

Similarly, even among the second-generation Asian Americans who were younger and did not have language barriers with their English-speaking neighbours, many of them described that their social contact with their neighbours was sparse. As one of them, Bentley Wong, informed me over a phone interview one evening, San Marino is "a private town" and an "indoor city." Bentley grew up in San Marino and for him, the city was simply a quiet and conservative place to hunker down that lacked any form of an active outdoor public space where residents could engage and connect with one another easily. Socializing in San Marino is largely behind closed doors in local clubs and private settings even though its residential community is small. Club membership thus plays a key role in structuring the social life in San Marino. Active local clubs in San Marino included the Chinese Club of San Marino, the parent-teacher associations (PTAs), City Club of San Marino, Rotary Club, and alumni clubs (e.g., University of Southern California or the National Taipei University). The clubs organized talks, fundraising for the public schools and civic services, charitable events, and social networking events for their members. Through these activities, these clubs indirectly provided opportunities for intercultural interaction within San Marino, albeit only among their members.

These interviews with residents show that social interaction in San Marino between social and cultural groups is highly selective, non-spontaneous, and circumscribed. During visits to community events such as Little League Baseball games, I observed that the Chinese sat with other Chinese, the Koreans sat with other Koreans, and White Americans with fellow White Americans; rarely was there a spontaneous mix. As one resident commented, "money can buy separation." It seemed that with resources in San Marino, one could select who one wants to meet or not meet.

Many of the "old time San Marino folks" interviewed told me that spontaneous encounters of neighbourly exchange in public spaces have become fewer because of the shifts in the cultural values and lifestyles of families who had moved in. Asian neighbours often kept to themselves and made less effort to build or sustain neighbourly relations. To better understand the reasons for these observations, I asked Chinese and Taiwanese residents regarding their thoughts about inter-ethnic friendships and interaction. Among those I asked was Nick Chang, a long-time resident who migrated from Taiwan. Nick's response was detailed and included the sentiments shared by other first-generation Asian immigrants. He explained his view on this tendency to hunker down. Note Nick's repeated use of this concept of "comfort" which I have italicized:

Basically, Number 1, it is my personality. I am low profile. If not necessary, I like to just keep everything in my family and myself. Second, my *comfort zone*. I feel *comfortable* speaking Chinese with my family. It does not necessarily mean that I am *uncomfortable* speaking English. I can speak English and communicate with other people.

But when you talk deep then I can't continue. That is the problem. Like when they talked about where they have come from and what they used to do. I cannot continue the conversation. They know that and then they change the topic. This is because of my background. I am brought up overseas in Taiwan. They received education here – like their childhood and school time, and their background – totally different. And the things we learn and the things they learn are totally different. The way they learn here is totally different from the way we learn from other countries. In Taiwan, we only receive. We don't give in the classroom. We just listen and try to learn. Here, you are encouraged to voice out your opinion from first grade.

As a first-generation here, your *comfort zone* is always going to a place where your social activities allow you to be involved in something that you can talk to your friends and your acquaintances in your mother language. Even if you educate other people and get mutual understanding of other cultures, you still don't feel *comfortable* to engage in social activities with the Caucasians, only if you need to ... but as far as your daily activity is concerned, you still want to be in your *comfort zone*. You want to engage in any activity that is Chinese speaking. You are more familiar with the people around you.

Ethnic and linguistic comfort zones act as double-edged swords in diverse settings. As cocoons of protection and refuge, these cultural enclaves provide needed companionship and resources to immigrants, but they also introduce new social and cultural boundaries to limit opportunities for intercultural interaction and learning. Take the case of Naomi, the resident who was actively involved in the PTA. A second-generation Asian American with Taiwanese roots, she experienced the impermeability of the linguistic comfort zone when trying to engage the new first-generation immigrants from China. Naomi had made multiple attempts to engage the new immigrant residents by offering to help them settle in. They liked Naomi and invited her to their private birthday parties and celebrations. At the social events, the new residents spoke in Mandarin Chinese, their comfort zone, and Naomi felt uncomfortable given her limited language capabilities in the language. She decided to stop pursuing these intercultural friendships further.

Scene Two: Central Long Beach

"I am Marteese Owens, and I am just trying to survive," a Black American man told me matter-of-factly. I had asked him about how he identified himself in the multi-ethnic and multinational environment of Central Long Beach. Marteese's response took me by surprise, as I was half expecting him to describe his age and ethnicity, like the other participants. Marteese, in his early thirties, was studying hard in the Mark Twain Neighbourhood Library for his college classes when I approached him for an interview. Marteese went on to explain the difficulties of competing with immigrants for jobs, particularly those from Mexico, even though he was born and raised in California.

Economic hardship and immigration grind at the formation of social life in this low-income neighbourhood.[12] In 2010, Central Long Beach had an immigrant concentration 1.4 times higher than the rest of the City of Long Beach and this concentration level remained in 2020.[13] The two major foreign nationalities in Central Long Beach were Mexicans (60 per cent) and Cambodians (18 per cent), while smaller groups of immigrants were from El Salvador, Honduras, Guatemala, the Philippines, Vietnam, Thailand, and China. With a median household income of about $33,000 in 2010 and $43,000 in 2020 (equivalent to about 60 per cent of Los Angeles County's median household income), Central Long Beach is a socio-economically impoverished area that receives direct food distribution up to twice a month. According to a local community organizer, these bi-monthly distributions serve about 500 Latino, Asian, and Black American families who live in a densely built residential area of run-down apartment blocks and subdivided single-family homes. Its poverty and density juxtapose sharply with the spacious single-family houses on well-kept, lush green boulevards, and clean urban sidewalks lined with chic cafes, boutiques, and spanking new loft apartment buildings, located nearby in East and Downtown Long Beach, respectively. See figure 3.5 for the location of Central Long Beach and refer to appendix 1.2 for the demographic information of Central Long Beach in 2010 and 2020.

"Long Beach is a segregated city," Randy Jones, a long-time resident of East Long Beach, concluded during our conversation. Before starting community organizing work in Central Long Beach, Randy, like many of his neighbours, had never had the need to visit Central Long Beach. He had not known about the demographic diversity outside East Long Beach that made the city one of the most diverse in the United States and one of two core concentrations of Cambodian diaspora in the nation.[14] A map in figure 3.6 of the distribution of demographic diversity in Long Beach illustrates the residential segregation in the city that Randy highlighted, where East Long Beach is mostly composed of White American residents while Central Long Beach has far fewer Whites and higher concentrations of Latinos, Asians, and Blacks.

Based on the cognitive mapping by the residents of Central Long Beach, the collective map analysis of routine geographies illustrates that few residents from Central Long Beach regularly venture out of the neighbourhood for the beaches and large parks in the rest of Long Beach. Instead, residents depend on the fast-food outlets, corner liquor stores, Asian markets, bodegas, video stores, and pocket neighbourhood parks strung along auto-centric East Anaheim Street for their daily needs and recreation. The social isolation and poverty concentration of the area are evident, causing several of its residents to pronounce it "a ghetto!"

According to long-time residents, the demographic transition from White homogeneity to ethnic heterogeneity began in the 1950s when Black Americans moved into the area. By the 1970s, Latino and Mexican immigrants arrived, followed closely by Cambodian refugees in the late 1970s. Between 1980 and 2000, Central Long Beach was notorious for its street shootings, active gang activities, blatant drug dealing, bullying, and street robbery. Older residents poignantly recalled attending a funeral every week during this span of twenty years. Central Long Beach was a "war zone" and a "tough neighbourhood" as violence became a daily affair between the Latinos, Cambodians, and Black Americans, according to the residents.

During the interview with long-time resident Kosal Sok, a Cambodian, he recounted his personal experience as a new Cambodian refugee settling in Central Long Beach after fleeing the horrors of genocide in the late 1970s. A seventeen-year-old youth then, Kosal had been repeatedly physically and verbally assaulted by other residents in the

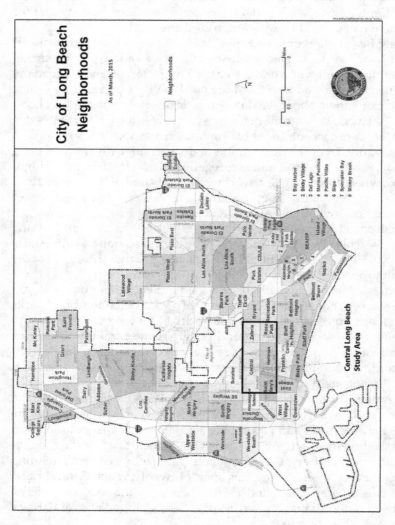

Figure 3.5. Location of Central Long Beach study area vis-à-vis other neighbourhoods in the City of Long Beach.

Adapted by author.[15]

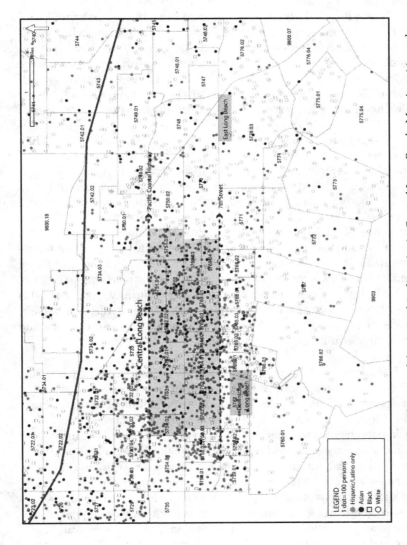

Figure 3.6. Map showing the distribution of demographic diversity in Central Long Beach. Map is not to scale.

Prepared by author using US Census data 2010 SF1 Table P5.

neighbourhood. He had been "kicked in [his] behind," while working at a store and then made fun of by non-Cambodians. He explained that Cambodians in the 1970s and 1980s had been "newcomers, the new kids on the block" and had not been welcomed by the residents in Central Long Beach. Refugee children had been bullied in school, and robbed of their new shoes and backpacks by other children while walking there. At school, the Cambodian children, who had spent years in refugee camps without education or an understanding of English, were placed in classes that matched their age but did not match their knowledge level. Many of the children negotiated the harsh daily realities by forming and joining gangs for protection. In his soft steady voice, Kosal explained vividly, "to them [the Cambodian children], they [the Cambodian gangs] protect me. Every problem, they take care of me. They buy me new shoes. When I am hungry, they buy hamburger for me." During the twenty years (1980 to 2000) of resettlement, many Cambodian youths either died in gang violence or were incarcerated, and many Cambodian teenage girls had babies outside of marriage. Kosal summed up poignantly, "this generation got lost."

The sense of isolation was exacerbated at home. Refugee parents, many of whom had escaped from rural Southeast Asia, were themselves unable to cope with urban life and the demands of adapting to a foreign culture and learning a new language, all while trying to recover from the trauma of genocide. Chenda So, a community organizer who worked closely with Cambodian refugees and was a refugee herself, described the survival struggles faced by Cambodian residents and the difficult inter-ethnic relations in the area:

The first generation, they [referring to the Cambodian refugees] came with their own historical genocide era and scarcity mentality. I have to look out for myself otherwise I am not going to make it. At the same time, the downside is that I learn to live with what little means I have. Instead of thinking: if I would go for higher education and advanced job, if I adapt and integrate into the mainstream society, I could enjoy the opportunity. They did not think that because I was told, realistically, you can only achieve the American Dream if you speak well, have a higher education, and get a job. If you have neither, you just barely make it out there. People who eat rice and soy sauce every week to get by and then Downtown where people throw food … When you throw in the people from the low-income community African American, Hispanic, and Cambodian, they start to have their own territory. Then you see gangs. And the idea was to defend themselves. The idea is to have my group and that is your group. And then what would that make to the society?

While inter-ethnic confrontations and gang violence had subsided due to multiple interventions by the police – such as gang intervention programs in schools, increased patrolling, and social support programs – the daily issues arising from commingled poverty, group divisions, and historical scars remained a bundle of challenges for many to negotiate.[16]

In 2000, a group of Cambodian businessmen who had businesses in the area but lived elsewhere in Los Angeles proposed to designate a part of Central Long Beach as "Cambodia Town." The project was resisted by the local residents, including the local Cambodians themselves who were strongly against the establishment of a singular cultural space in the multi-ethnic environment. Social tensions reverberated in the neighbourhood as fears of violent inter-ethnic reprisals pervaded and threatened to jeopardize the fragile peace achieved among the different social and cultural groups. After more than a decade of a tortuous process filled with disagreements between multiple stakeholders, a one-mile-long strip (about 1.6 kilometres) along East Anaheim Street (the main thoroughfare of Central Long Beach) was finally officially identified as the "Cambodia Town Cultural District" in 2011 (Chan 2013a). See figure 3.7 for pictures of Central Long Beach.

Ethnicity as an Organizing Principle in Social Space

Residents and community organizers experienced a heightened awareness of ethnicity as an elemental differentiation and a daily organizing principle of social relations in Central Long Beach. Randy Jones observed that the people whom he worked with in Central Long Beach were "more honest" with race as compared to those in other parts of the city. Randy gave the example of the use of ethnic categories in daily communication, such as "Who is the White guy (or that Asian person) you were talking to?" These common expressions reflect the pragmatic civility used by residents to navigate diversity. However, the daily dose of differentiation based on visual "categoric knowing" (Lofland 1973) has also perpetuated ethnic profiling and stereotyping among its residents. When the residents were asked during the interviews if understanding was lacking among the different ethnic and nationality groups, more than half felt that there was insufficient understanding because of the lack of engagement with, learning of, and adaptation to one another's cultures, with many of the residents citing stereotyping as a chronic condition in the neighbourhood.

Civil discord in Central Long Beach can quickly turn into inter-ethnic discord because of the common use of stereotypes to attribute the problems arising from individual actions as a group behaviour trait. For

Figure 3.7. Everyday landscape in Central Long Beach. (*top left*) The main commercial thoroughfare of Central Long Beach along East Anaheim Street. (*top right*) Children playing along a sidewalk on a Saturday late morning. (*bottom left*) Latino mother with her children walking along East Anaheim Street on a weekday noon. (*bottom right*) Teens walking home after school along Cherry Avenue.

Taken by author in 2011 and 2012.

example, the shortage of street parking space in Central Long Beach is often blamed on the Mexicans, the majority population who are seen as hoarding precious street space by not parking their cars properly. Frequent agitations like this between neighbours of different ethnic groups have led to several heated arguments and even fights over the rights to use certain spaces. Alina Daniels, a Black American resident in her early twenties who grew up in Central Long Beach, openly shared her thoughts on inter-ethnic relations in the area:

To be honest, I really don't keep my eyes on the Cambodian people and Asian people coz they are really not in conflict of the international racial part we have going on in our building, in our surrounding areas. It is majority Mexicans and Blacks. The Cambodian people and Asians, you know, they keep to themselves.

Alina expressed a common perception of inter-ethnic relations held by residents and community organizers. Apart from East Anaheim Street where many Cambodian businesses are located and the indoor Mark Twain Neighbourhood Library where the Asian presence is visibly felt, Asians are often described by others as "invisible." Compared to other ethnicities, Asians are, in fact, rarely seen walking on the streets or using the neighbourhood outdoor public spaces. During the interviews, Asian residents spoke about their preference to travel by car as much as possible within the neighbourhood in order to minimize their exposure to street violence. For example, long-time resident Munny Ly, originally from Cambodia, travelled by car everywhere, even to the grocery shop nearby, to avoid walking on streets that she felt were unsafe. For her daily morning walk, she would only go on a circumscribed circuit of a two- to three-block radius around her house because of personal safety concerns. In my interview with Chenda So, a community organizer in the area, she explained the reasons behind the invisibility of Asians in the neighbourhood. From her point of view, it was an outcome of both the cultural preferences and environmental circumstances of Central Long Beach:

These people [Cambodians] are very friendly people. When they trust you, they are very loyal ... The negative part of it is that if you don't make initiative and connect ... Asians don't do networking. They don't go to the neighbourhood park, and they know little about each other. They don't talk to each other, and they keep to themselves. They tend to close the door and shut the door and so that they don't get help ... The kids learn how to survive in the tough neighbourhood ... When you hit a little more crime area, you keep to yourselves more. So what kind of relationship are you going to develop? You are not going to have a free environment where you can trust each other when you don't even talk to each other.

On the other end of the visibility scale is the palpable tension between Black Americans and Latinos. Competition for jobs between these two ethnic groups, in addition to the gang rivalry between them, has made these groups very visible to each other and to others. Black American residents like Marteese and Calvin Jenkins described how they were being "run over" by Mexicans looking for employment and housing. Common racial slurs and entrenched stereotypes are exchanged between Black American and Latino youths to provoke discord and incite fights that have in turn, further increased the hostile feelings between the groups. Ben Rodriguez, a Mexican American in his twenties, expressed his frustration with the lack of intercultural relations in Central Long Beach:

> There is not enough learning … like our backgrounds and where we came from. What we had to do, what we are now … People just go on discriminating and do all that stuff. They don't understand it. They judge first. They judge people … They don't say out loud sometimes, but you can tell they are saying it. It is just like the look they give you. I understand those things.

In addition, friendships between Mexicans and Black Americans are particularly difficult to form and sustain in Central Long Beach, according to Marteese who grew up in the neighbourhood. In a low whisper, he explained:

> Everybody grew up with everybody. I used to have a Mexican friend, and I used to go to his house to eat burritos, tacos, and carne asada and all kinds of different things … We know of each other but just tend to outgrow it. You tend to want to go back to your own culture that you know more of … it becomes sort of a hatred … Let's say I am Black, he is Mexican. I know my cousin and probably how jealous he would be because we grew up together. I don't want him to do anything bad to my Mexican friend. So I kind of pull away because I don't want him to do anything to hurt him [his Mexican friend] or get himself [his cousin] into trouble. It is a dead end. So I am trying not to get anybody hurt and so I pull away. And probably like my cousin is in a gang and my Mexican friend is in a gang. You don't want to be the one who lights that match … There is always somebody trying to break up the group. So we tend to stay away from the groups because there is always somebody trying to break up the group, so why try? When we finally having a good time. Now the police is trying to mess it up. Blacks and Asians don't fight, even the Samoans. For some reason, I don't know why, it always affects the Mexicans. [He lowers his voice.] The police always influence the Mexicans to turn their backs on us.

Parallel Practices in Shared Space

Street violence and fear of danger have created a hostile environment for social life in Central Long Beach. Daily social interaction between neighbours of different ethnic and nationality groups according to residents is limited to familiar neighbours. Even with these neighbours, the exchange is at best, composed of fleeting hi-bye greetings exchanged on the sidewalks. Sometimes, lasting interpersonal relations might form between next-door neighbours, but it is rare between those of different ethnicities and nationalities. Recounting the multiple disappointments that he had experienced in inter-ethnic encounters, John Turner, a Black American man in his early thirties, told me in a cynical tone,

> As far as meeting people around here, it's kind of normal in a way that certain people like to stick to their own people. The ghetto is kind of like that. This part of town is rather ghetto. This is the way it goes. People want to stick to their own race and people they already know. They are not so much into meeting new people. For example, there is a pool hall down the street where a lot of guys play pool. The guys are mostly Asian. I try to go there a few times, but I find that I am not really welcomed ... They are all Asians and they are all doing their own Asian thing. I am kind of an outsider. 'You don't know our food, you don't know our custom. We don't want to teach you.' Just not really welcomed there. It is one of the things you just have to accept ... if I go to another pool hall like Lakewood, I will be accepted. This pool hall here is small, and there is a small group of guys who go there and they don't care for outsiders.

John's experience is not unique, as many participants have related that social life in Central Long Beach is usually undertaken within a social group and not across groups. The participants commonly used expressions such as "sticking to their own," "sticking to themselves," "minding their own business," and "everyone for himself" to describe the social atmosphere in the area – not unlike Putnam's (2007) findings of low trust and hunkering down in diverse neighbourhoods. In Central Long Beach, ethnic boundaries are enforced between groups and individuals by the process of inclusion and exclusion from certain activities and public places (Barth [1969] 1998). Ethnic boundaries fragment the social space of the locale so that social life is ordered along group lines, and this condition further diminishes the occasions and possibilities to develop relations across groups as living and recreation arrangements become segregated. According to several long-time residents in Central Long Beach, one frequent practice is found in apartment buildings, where apartment managers tend to favour tenants who share their ethnicity or nationality and thus, over time, these buildings become vertical ethnic or national enclaves.

The practice of social life along ethnic boundaries is also obvious along East Anaheim Street, the main thoroughfare threading east to west through Central Long Beach. East Anaheim Street is frequented for its public amenities, such as the library and park, as well as the

corner stores, grocery shops, restaurants, and neighbourhood businesses owned and operated by a mix of Latino, Cambodian, and Korean
businessmen. Visibly, it is a space of intermixing and permeability
where different shops and clienteles engage in their daily activities.
However, a closer study of the routine patterns through the cognitive
mapping interview (shown in figure 3.8) reveals that Cambodian residents visit Seng Heng Market for Asian goods while Latino residents
visit the Fifteenth Street Market. Thus, social interaction is mostly fleeting between groups and at best a "routinized" co-presence in Lofland's
(1998) descriptive categorizing of relations in the public realm. There
are few to no places other than the neighbourhood park and library
within Central Long Beach where residents of different ethnic groups
can gather, linger, and possibly intermingle. Randy Jones shared with
candour his observation about the social life of East Anaheim Street:

MYSELF: Do the different ethnicities mix?
RANDY: They don't mix.
MYSELF: Why?
RANDY: Fear and tradition! [*pausing for a few seconds*] People are afraid of
    people who don't look like them. And there are a lot of gang activities
    that probably got a lot of people spooked. A Black American guy
    encounters a group of Latino guys. He is not going to start a wonderful
    multicultural encounter. And there are also not so many opportunities
    for people to mix … The Latinos go to Latino groceries and they only
    speak Spanish. The Cambodians go to the Cambodian groceries, the Seng
    Heng Market, and the Kim Long Market where they speak Khmer. It's
    diverse but segregated.

I observed that recreation activities tended to be practiced within a single
ethnic group. My observation was verified by long-time Latino and Black
American residents. Asian youths gathered regularly at street corners
along East Anaheim Street with their skateboards. Latinos played soccer,
while basketball appeared to be played only by groups of Black American
men in the parks. However, I was told that there had been occasions in the
past when Black Americans and Latinos played basketball together and
compete in a friendly game but the memories of those times were faint.
According to both the Black American and Latino participants, the reason
for the rarity of mixed teams was because Latinos often speak only Spanish among themselves while playing, which made it difficult to include
Black Americans in the same team. Similar patterns of minimal socialization and parallel existence (Cantle 2005) were also observed among the
elderly at McBride Recreation Center. Black American men played pool

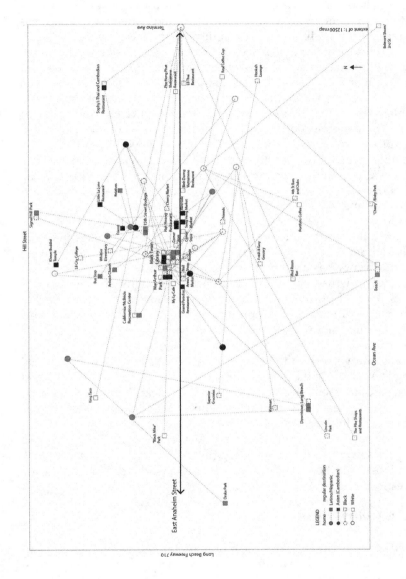

Figure 3.8. Mapping routine geographies of Central Long Beach participants (N=19). Map is not to scale.

Prepared by author.

in one room while Cambodian men gathered in another room to chat. The Cambodian women socialized almost exclusively with each other during dance lessons and lunch. The climate of cultural enclaves in this community centre demonstrated that coexistence in the neighbourhood was structured by ethnic, linguistic, and at times, national similarities.

There are very few public places in Central Long Beach that residents can freely and safely access without having feelings of exclusion and fear of danger. The Mark Twain Neighbourhood Library is one such place, where glimpses of intercultural learning and relations are made possible by the activities and operations of the facility. Built and opened in 2007, the neighbourhood library has become a safe house and refuge space for the residents in the area. The library is very popular across all demographic groups. Mothers bring their children to the library for free after-school tuition, and to simply hang out in a safe space in the afternoon because their homes are often overcrowded and not conducive to learning. The Mark Twain Neighbourhood Library recognizes the legitimacy and presence of the different linguistic groups present in the area. It houses the largest collection of Khmer language books in the United States, in addition to Spanish and English books. Furthermore, the library has a team of multilingual staff who can speak Spanish, Tagalog, Vietnamese, and Khmer that helps many of the social and cultural groups feel welcomed. The library is also the host of the weekly Khmer language class and community meetings, and has a notice board that posts information of major events in the community.

According to the librarians and the residents, the library is a central community space in the neighbourhood. It is a place where social mixing is made possible because residents from all different social and cultural groups are welcomed. Its openness, cleanliness, and sense of physical safety relative to the surroundings, also make the space inviting for intermingling. According to the head librarian, the library is a "neutral territory" where residents feel a sense of familiarity and belonging. Given the active gang-tagging activities in the area, the lack of vandalism on its building, save for one episode, demonstrated the tacit respect it has in the community.

Scene 3: Mid-Wilshire

Located at approximately the mid-point of the sixteen-mile-long (about twenty-six kilometres) Wilshire Boulevard that runs from Downtown Los Angeles to the Pacific Ocean, the Mid-Wilshire locale is a collection of neighbourhoods, and cultural and ethnic towns, such as Koreatown, Little Bangladesh, and the Salvadoran district.[17] Poverty is juxtaposed with wealth in this culturally eclectic area of the city. Run-down and

Figure 3.9. Myriad of physical, social, and cultural diversities in Mid-Wilshire. (*top*) manicured sidewalks and outdoor leisure dining in western Mid-Wilshire and (*bottom*) crowded and busy streets in the eastern half.

Taken by author in 2012.

densely built rental apartments stand alongside private condominiums that quickly transition into large mansions with manicured grounds in Hancock Park. Within a short 2.5 miles (about four kilometres), the annual median household income in 2010 ranged between US$26,000 in the eastern end near Downtown Los Angeles and US$130,000 in the western end bordering the famously rich Beverly Hills. This income disparity remained unchanged in 2020.[18] Home to multiple generations of immigrants from over 100 nations including Guatemala, El Salvador, Mexico, Korea, the Philippines, and Bangladesh, as well as a mix of Hassidic Jews and White Americans, Mid-Wilshire is a multi-ethnic, multinational, and multilingual globalizing area. Figure 3.9 shows the different landscapes found in Mid-Wilshire. See appendix 1.3 for the demographic information of Mid-Wilshire in 2010 and 2020.

In explaining the geography of Mid-Wilshire to me, Marcus Kenny, the municipal officer responsible for its urban development, identified "the big dividing line" of the neighbourhood. According to Marcus, Wilton Place is a busy two-lane roadway that splits Mid-Wilshire spatially and socially into two parts – the western side with more privately owned residential properties and less density, and the eastern side where densely built mid-rise rental apartments are interspersed with commercial corridors. Demographically, the western part is composed of a largely White population interspersed with Asian homeowners, while the eastern part comprises predominantly Latino and Asian immigrant tenants. See figure 3.10 for a map showing the ethnic distribution in the area according to US Census 2010. It illustrates quite starkly the difference in density and ethnic diversity between the two halves of Mid-Wilshire.

Marcus informed me that neighbourhood activists proposed to split the Wilshire community along Wilton Place in 2001 but city hall swiftly rejected the proposal. See figure 3.11 for the Generalized Land Use plan of the Wilshire area, showing Wilton Place as the big divider in terms of land use, where the west of Wilton Place is composed of single-family homes and open space, while to the east, multiple family apartments and commercial use characterized the area.

This split remained salient as Western Avenue (the major artery adjacent and parallel to Wilton Place) functions as the political and institutional boundary between the Greater Wilshire Neighbourhood Council that represents the interests of the high-income "westerners" and the Wilshire Center-Koreatown Neighbourhood Council that represents the lower-income "easterners." When taken as a whole landscape, as Marcus explained, Mid-Wilshire appears visually and socially "mixed," but a closer look quickly reveals that the kinds of sociocultural diversity, buildings, and urban forms are in reality polarized between the east and west of Wilton Place. For example, the western part has quite a number of historic preservation overlay zones that seek to preserve structures and landscapes of historic value to the city, such as Hancock Park and Windsor Square. The planning intention for the area is slow growth and improve stability for the neighbourhoods. In comparison, the plans for eastern Mid-Wilshire seek to improve the economic opportunities through improvements made to the built environment, for example through the Wilshire Center Business Improvement District and the East Hollywood Targeted Neighbourhood Initiative. Overall, city hall conceives the eastern part as an area that needs intervention in order to ensure constant development for economic growth, while it regards the western half as a place that needs help to fend off change and redevelopment because of its cultural and social significance.

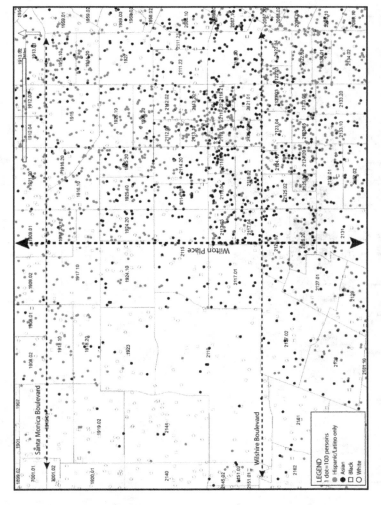

Figure 3.10. Map showing two distinct spatial and demographic identities in Mid-Wilshire along Wilton Place. Map is not to scale.

Prepared by author using the US Census data 2010 SF1 Table P5.

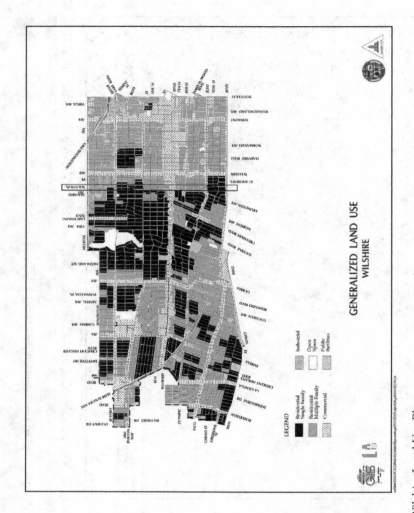

Figure 3.11. Wilshire Land Use Plan

Adapted to greyscale by author

Contested Space and Cultural Enclaves

A locale of multiple immigrant group settlements, Mid-Wilshire is a place of overlapping interests. The historic importance, political centrality, density, and demographic complexity of Mid-Wilshire in the context of the City of Los Angeles make it a highly contested area where different social and cultural groups tussle for recognition and symbolic claim to a space they can call their own. For example, advocates of Koreatown, supporters of Little Bangladesh, and Salvadoran business owners and immigrants have fought over the rightful boundaries of their community space (Jang 2009). In another example, the exclusive Wilshire Country Club had a disagreement with the former Community Redevelopment Authority (CRA) over a new street landscape with Korean cultural motifs. According to a municipal officer familiar with the case, the proposal unleashed strong retaliation from the elite country club, who said, "This is not Koreatown!" In yet another case during the 2010 redistricting of council districts, community organizers and residents of Koreatown lobbied hard to unify the representation of their interests from four council districts to one. However, pushback from other interest groups in the area to keep the status quo were equally strong.

Unlike the social and cultural coherence that characterized more traditional immigrant ethnic enclaves of Chinatown or Little Italy, ethnic towns in Mid-Wilshire like Koreatown and Little Bangladesh comprise an eclectic mix of national and ethnic identities within their borders. At the time of research in 2011–12, Koreatown's Eighth Street had more Hispanic business services than Korean ones, while Sixth Street had a concentration of Korean restaurants and businesses. On other streets, one would easily find Korean businesses interspersed with Mexican and Salvadoran shops. This pattern of mix was also evident in nearby Little Bangladesh, a ten-block linear commercial corridor whose boundaries partially overlapped with Koreatown. Bangladeshi businesses (restaurants, grocery shops, gas stations) occupied only a couple of street corners, while Korean businesses and Hispanic retail shops and restaurants lined the street in between them. In comparison to the modest presence of Bangladeshi and Salvadoran culture, the Korean cultural district was more visibly distinct, as large signboards in Korean script were found outside and inside the Koreatown official boundaries, and the overall size of the district was much larger to begin with.

Although the cultural landscape of Mid-Wilshire appears diverse, the extensive variety of businesses, amenities, and services, such as specialty supermarkets, churches, restaurants, car-repair shops, and hairdressers

catering to specific ethnic clientele fragments the landscape into multiple ethnic bubbles – self-sufficient cultural environments that have little
interaction with each other. Hannah Youn, a community organizer and
resident of Mid-Wilshire, explained that ethnic bubbles were formed
because the easy availability of ethnic businesses made it convenient
for residents to patronize shops where they could transact comfortably
in Spanish or Korean, so there was little incentive or necessity to cross
cultural borders for these daily spatial practices. Thus, there was little
permeability between the bubbles. Furthermore, residential sorting
along lines of ethnicity or nationality is quite common. According to
residents, some street blocks in western Mid-Wilshire were predominantly occupied by Hassidic Jews, and there were vertical enclaves
in eastern Mid-Wilshire that comprise Hispanic or Korean residents
in apartment blocks where communication was usually in Spanish or
Korean. In one such condominium that I visited for an interview, the
building had notices in Korean only. The resident whom I interviewed
verified that her neighbours were mostly, if not all, Koreans or Korean
Americans, and there was perhaps only one non-Korean living in the
condominium.

During the fieldwork, I met some residents who had lived in both
the western and eastern parts of Mid-Wilshire. I asked them to contrast their experiences of living in both areas. Two of them were young
second-generation Hispanic Americans (Luciana Garcia and Eileen
Corez) who had grew up in western Mid-Wilshire and then relocated to
eastern Mid-Wilshire with their families. Luciana and Eileen felt that the
ethnic bubbles in eastern Mid-Wilshire were larger and less penetrable
than in western Mid-Wilshire. Eileen Corez, an articulate and confident
second-generation Hispanic-Black American, felt that there were more
opportunities for intercultural learning in western Mid-Wilshire than
eastern Mid-Wilshire. Eileen described her experience as she referred
to the map of Mid-Wilshire:

> Where there are a lot of Hispanics, they don't get the opportunity. It could
> be to the point that they are so solid until that they are used to it and not
> want to try new things. Over here [referring to the western Mid-Wilshire] it
> is more open. It is more closed-off over here [referring to the eastern Mid-
> Wilshire] because there is more of one race.

Luciana, a candid and cheerful young second-generation Mexican
American who used the Wilshire Branch Library regularly to do project research for her classes, vividly described her experiences walking from the neighbourhood library in western Mid-Wilshire to her

home in eastern Mid-Wilshire. She said, while pointing to her cognitive map,

> When I am here at Oxford Avenue, I feel like I am in my own community ... From there I don't worry. Just a couple more blocks I am home ... They [referring to the Koreans] are all right there. They work. They have their own shops and restaurants. When you come to the Latino community, that's home. When I am there, I am calm. Nothing is going to happen to me here because I am near my community ... We don't see Koreans and Japanese as danger for us, and probably the Black Americans, you fear. Sometimes, oh damn, they are here! When you see a group of guys hanging around, you think like something might happen to me. With the White community, we actually never have communication with them. People start judging – "Those White people don't like us."

Drawing on these everyday experiences of living in diversity, the ethnic bubble is not simply an abstract concept but has material presence as it provokes physiological fear and threat for an individual living in polarized and sorted conditions of diversity.

Relational Web in Density and Diversity

Particularly among the non-White residents, their experiences of social life in Mid-Wilshire spoke of thin neighbourly relations and a hunker-down posture even within social groups. During one of my afternoon visits to the neighbourhood library, I interviewed Chloe Castillo, a friendly, open, and bubbly Filipino immigrant resident in her thirties. Chloe lived in an apartment building with a mix of Filipinos and Mexicans in eastern Mid-Wilshire, and when I asked her about the social life in her apartment block and the neighbourhood, Chloe expressed a sudden negativity. She spoke about her cumulative bad experiences living with inconsiderate and uncooperative Mexican neighbours and confided that she could not wait to move out of the neighbourhood to nicer suburban parts of Los Angeles where her cousins live. Chloe said:

> Most of the people living around here are like Mexicans [*lowering her voice*] and Filipinos. We don't like and we don't talk. We have nothing to do with each other. They [Mexicans] just ignore us. They don't help. Sometimes you need to ask a favour to move the car a little so you can park.

Mi Young, a soft-spoken South Korean immigrant who lived in another apartment building in eastern Mid-Wilshire, shared similar sentiments. She described the contact with her Korean and Hispanic neighbours as composed of fleeting encounters and contrasted it with her experience of living in suburban Orange County for some years, where she felt that routine neighbourly relations were overall friendlier with her White neighbours. However, several incidents of discrimination experienced by her son in the playground motivated her to relocate to Mid-Wilshire with the hope that its demographic diversity would expand the possibilities for social inclusion. However, as Mi Young explained, high-density apartment living was not conducive to relationship building. She said:

> Mostly in apartment complex, they don't know each other well. They just close their door and don't know who lives next door ... people are individualistic, and we don't interact too much. Here, Hispanics and Koreans ignore each other. They don't say hi. You can tell that they are not happy to live with each other.

Does living in high-density environments make it hard to form good social relations with neighbours in diversity? Speaking to residents who lived in the relatively more spacious environments western Mid-Wilshire, I discovered that the experience of isolation was not unique to the residents in apartments. Residents living in the lower-density western Mid-Wilshire also spoke about a lack of neighbourly contact.

Nancy Lau, a long-term Chinese American resident in western Mid-Wilshire, owned a large single-family home in one of the green and lush streets of Hancock Park, also known as "Candyland" among the local residents because of its large houses and well-kept lawns. Nancy had raised her children in the neighbourhood and knew the area very well, having grown up in Los Angeles and worked in the city all her life. She told me that the only streets that she had observed to have active neighbourly exchange and community activities are those streets where many large Hassidic Jewish families resided. On her street, there was a mix of ethnicities and nationalities, but neighbourly contact was infrequent and social relations were composed of what Lofland (1973) theorized as "categoric knowing" – a superficial knowing of another's ethnicity, occupation, and other demographic life stage characteristics. Nancy said,

> It is not real connected in terms of the concept of neighbourhood. It is basically a place to live and you do your business. It is not a traditional neighbourhood where you actually know your neighbours ... It is not a neighbourhood where children played with each other – that kind of

neighbourhood ... Our children went to private schools. Their connection was at school ... Like Sarah across the street, her children go to a Jewish school. His child [*pointing to her next-door neighbour*] goes to school in Brentwood. None of the children really know each other or play with each other. We know the families and names and maybe what they do. There is really no interaction in terms of neighbourhood, community, recreation. They don't even play in the same soccer team. You do your own thing, you live here but your activities are elsewhere. So, it is not a traditional neighbourhood.

In contrast to these tepid relations, I observed that the intercultural experiences of White American residents were markedly different from those of the non-White residents whom I interviewed. Jenny Fellow, a resident in eastern Mid-Wilshire for over twenty years, lived in a small apartment building and enjoyed the company of her Filipino neighbours. Jenny felt that Mid-Wilshire is multicultural – she sensed an overall respect for cultural differences here, based on her good relationships with her neighbours and her frequent sightings of people of different ethnicities in her neighbourhood chatting with each other while waiting to pick up their children from school. At another interview, I met Larry Gans, a western Mid-Wilshire resident for over two decades, who had neighbours from different ethnic and national origins on his street of single-family homes. He recounted the reciprocity and friendliness among his neighbours, who would readily help each other out, be it lending tools or helping with minor home projects. Then, there was Alison Haynes, a long-time resident who lived in the transition zone between western and eastern Mid-Wilshire. Alison and her husband had chosen to live in Mid-Wilshire for its sociocultural diversity, where daily life for her comprised interesting and enlightening intercultural encounters with her Persian, German, and Korean neighbours. Alison spoke about how she had enjoyed the opportunities to patronize a Korean-French bakery, shop at a local Persian grocer, and frequent one of the biggest Korean supermarkets across the street. The accounts of White American residents exhibited that social life in diversity was sometimes capable of *intimate-secondary* relationships of deeper and more lasting relations between neighbours, even if the overall relational web in Mid-Wilshire comprised mostly of *routinized* quick hi-bye and short acknowledgments (Lofland 1998).

The White residents' overwhelmingly positive responses to intercultural interaction with neighbours of different ethnicity and nationality could have been the outcome of self-selection – they had sought to live among multiple ethnicities and nationalities because they enjoyed sociocultural diversity to begin with or had acquired the enjoyment and competence to live in diversity. There was also a possibility that my presence as a non-White interviewer might have influenced how positively the participants responded about diversity in order not to appear impolite.

In Mid-Wilshire, there were also incidences of what Lofland (1998) termed as *quasi-primary* relationships among strangers in the public spaces. These encounters lasted longer than a typical routinized greeting between strangers, such as conversations that may involve exchange of personal information. Resident Mark Adams termed these encounters as "LA small talk," when familiar and complete strangers in Mid-Wilshire become adept at a ritualistic social practice of asking common questions about strangers with the purpose to access information about another, to satisfy curiosity about difference in social complexity, and according to Mark, to diffuse tension arising from differences. According to other residents in Mid-Wilshire, "LA small talk" was a process that was highly personal and reliant on visual cues. Making eye contact and giving a little nod, and complimenting another's accessories such as shoes, bags, or even hairstyle, clothes, dogs, children, etc., were ways to establish contact between strangers in Mid-Wilshire.

In addition, immigrant participants shared that "LA small talk" for them always included a question of "Where are you from?" which they felt was a result of their accented non-American English. "LA small talk" between familiar strangers, such as neighbours living on the same street, usually comprises brief conversations about the weather, neighbourhood events, weekend plans, or each other's children. These conversations simulated moments of quasi-primary relations between people in a "world of strangers" borrowing from Lofland's (1973) conceptualization of social life in the city. Lucas Alvarado, a well-mannered and driven second-generation Mexican American who had grown up in Mid-Wilshire and attended a local college in the area, described vividly how negative tensions arising from mutual stereotyping between Hispanics and Black Americans were skillfully smoothed over through "LA small talk" on the sidewalk along his street block.

MYSELF: What do you talk about?
LUCAS: Just about the day, like small chit-chat on the porch, in the front of the building.
MYSELF: Do the people live in your building?
LUCAS: They are pretty young, a couple of young people around my age, less than thirty [who live] in my building and the building in front of me. It is not really much of like talking to get to know each other. Talking to spare

some time because we are there ... It is basically just saying "hi." It is kinda like, "What you up to?" We talk something about the block, like people, stuff like that, what is happening on the sidewalk. The other day, a girl got mistreated by her boyfriend. We are just talking about that. It is not like we are sharing stories or anything like that. It is kind of like very small talk.

From this perspective, "LA small talk" shares traits with sociologist Elijah Anderson's (1990 and 1999) writings about street wisdom that is necessary to navigate racially mixed environments as well as to secure safety in these spaces. Based on his later field research of poor African American inner-city neighbourhoods, Anderson (1999) further refined the conceptualization of street wisdom as a "code of the street." According to Anderson (1999), a "code of the street" is a set of tacit and informal rules that governs street behaviour, and these rules have been developed by residents to negotiate the street violence and adapt to a lack of formal institutions in regulating personal safety. In similar ways, "LA small talk" is a streetwise social device or code to navigate a complex social environment with perpetual strangers from different origins by gathering information about them, forming an ephemeral communication commons to diffuse tension between strangers, and even conceivably engaging in an interpersonal face-to-face contact to feel human and known.[19]

Apart from the sidewalks close to home where neighbours and familiar strangers may meet and briefly encounter each other, Mid-Wilshire has few public places where people of different ethnicities and nationalities can converge and interact. According to the routine destinations of residents within Mid-Wilshire, the popular places in Mid-Wilshire are social enclaves. Larchmont Village, an upscale main street lined with cafes and boutiques in western Mid-Wilshire, serves a predominantly White and Korean clientele, while Koreatown restaurants and cafes are largely patronized by Asians. In these locations, an individual from a different social group can be made to feel like they are trespassing. Longer-than-usual gazes or questioning looks from clients in the shops make visiting these places uncomfortable and exclusionary for some. Liz Joo, the Korean American resident highlighted in chapter 1 who grew up in Los Angeles, lamented the lack of collective life in Mid-Wilshire as compared to the different American cities she had lived in and other areas in Los Angeles:

There is no contact because there is no space to. Where do you go where you see other people? There are not many restaurants. It is not like you go to someone's house when you don't know them. Larchmont, which is one-block long, is the only place you kinda can walk around and eat ... where would you bump into somebody? ... If it was a culture where a lot of people are walking, let's say San Francisco or New York, maybe you would see many of your neighbours' faces. I want to see my neighbours sometimes but I can go like four, five months not seeing my next-door neighbour, coz if you miss them in the elevator, you are not going to see him, and I don't know when he goes to work. I can go for a long time without seeing any of my neighbours. There are just no opportunities. Maybe if there is a gym. If there is some place like you are doing something.

A closer study of the routine destinations of residents revealed that there were only two prominent public places in Mid-Wilshire. Burns Park and Wilshire Branch Library were public spaces that residents of different social and cultural groups from eastern and western Mid-Wilshire regularly visited. According to the interviews with residents, these places were unique because of the open atmosphere for intermingling and experiences of positive encounters across social and cultural differences. In fact, at these locations, particularly in Burns Park, residents had experienced good conversations with familiar and complete strangers from different social and cultural groups amounting to a blend of *quasi-primary* and *intimate-secondary* relationships, following Lofland's (1998) categorization.

The attraction of Burns Park was steady, long-lasting, and far-reaching among residents of Mid-Wilshire. Some residents with children living in eastern Mid-Wilshire travelled by foot and/or by car to use the park, even though there were public parks with playgrounds nearer to them. Residents regarded Burns Park as a safe and clean public place, especially important for those with children. At the park, children of different ethnicities and socio-economic backgrounds had the opportunity to socialize and play together, while parents and nannies intermingled. On many occasions, I watched parents and nannies (Korean, Hispanic, White, and Filipino) encouraging their children to befriend other children of another ethnicity and to share their toys and snacks with each other. It was also a space for parents to exchange parenting advice and give support, especially when their children played with each other. Users of the park also indicated that they liked the tacit sense of community and responsibility shared by the adults of all ethnicities to look out for the safety of the children. Burns Park created fond childhood memories for many, including two participants, who recounted memorable experiences they had in the park over the years with friends and families. One Hispanic resident who had moved out of the neighbourhood even made it a point to bring her toddler to the playground because she had enjoyed the place when growing up in Mid-Wilshire. A similar sense of a temporal community was also apparent in the Wilshire Branch Library. The users whom I interviewed spoke about the familiarity and trust they had developed with fellow library users, who often were regarded as familiar strangers. The participants separately spoke about occasions when they had helped to look after each other's belongings when someone needed to step out to make a call or visit the restroom.

Concluding Thoughts: Multivalent Diversities in Los Angeles

The three scenes of cheek by jowl diversity in Los Angeles offer glimpses of the "throwntogetherness" (Massey 2005) of values, practices, and ideologies common in the everyday life of a resident in the demographically and culturally diverse neighbourhoods of the metropolis. These snapshots of different kinds of sociocultural diversity point to the different tactics of negotiation that have been developed to respond to the different set of wider environmental, cultural, social, and economic conditions that interact and influence the residents and their behaviours. For example, the influence of "LA small talk" appears to be prevalent in Mid-Wilshire but not in the other two locales, even though San Marino and Central Long Beach are located in Los Angeles County. How do we explain for this difference? Is it an outcome of the interaction of contextual factors specific to the diversity of the locale that makes "LA small talk" more salient and cognizant among its residents in one locale than the other? Or, is its prevalence in Mid-Wilshire an indication of the necessity for a highly developed social lubricant tactic to negotiate the greater density and intensity of diversity found there? In the following chapters of the book (chapters 4, 5, and 6), I will delve deeper into the similarities and differences of the experience of living in diversity through an analysis of tensions arising from how social relations are organized in space. This discussion will explain the constraints and potentialities of diversity for collective life for the city, and serve as the foundation upon which to explore the possibility of developing intercultural public space design principles in chapter 7.

4   Tensions in Diversity

LA can be viewed like a powder-keg always waiting for one little match to blow up into something big because we do have many different cultures here, a lot of history in the city, a lot of tensions in the past that can crop up again. Most of the residents were not here during the 1992 riots. It does not mean much to them.

Sergeant Caleb Torres, Los Angeles Police Department
police officer in Mid-Wilshire

The discussion of social tensions arising from cultural difference is not new. During the post-Second World War years, UNESCO embarked on the Tensions Project (entitled "Tensions Crucial to Peace") to gain greater knowledge about the dynamics of intergroup relations in a bid to improve and secure international understanding.[1] Studying social tensions was relevant because as Director of the Tensions Project Robert C. Angell wrote, social tension is an early indicator of potential conflict between groups or persons and hostile attitudes between people (Angell 1950). Thus, social tension can be a precursor of overt social conflict or what German American sociologist Louis Wirth (1949, 142) referred to as "conflict potential" or "latent conflict" in his comments on the Resolution of the Economic and Social Council on the Prevention of Discrimination and the Protection of Minorities:

> The terms "intergroup tension" and "intergroup conflict" have sometimes been used interchangeably. It would seem useful, however, to distinguish between them by using the concept tension to refer to the *conflict potential*, and to speak of conflict when overt action takes place. Tension is, thus, *latent conflict* and may, among other ways, be resolved by conflict. We *infer* the existence of tension from the presence of attitudes or behaviour which leads us to believe that overt conflict may be in the offing, although not all

tensions result in open conflict … It should be recognized, of course, that tensions do not inevitably lead to overt conflict, but are a pre-condition of it. All life involves some tension.

Wirth highlighted that tensions often arise in competition for resources and are prevalent in social changes when groups are subject to unequal treatment, resulting in differentiated well-being. However, Wirth (1949, 142) explained that tensions are not always destructive, and they may be constructive by having "an organizing and integrating effect upon groups and societies." Similarly, conflict should not be viewed only as "disruptive, dissociating and dysfunctional," according to Lewis A. Coser in his seminal book *The Functions of Social Conflict* (1956, 21). Instead, conflict can function as a mechanism that helps to stabilize society over time. Conflict is destructive in circumstances where social structure is rigid and leaves no room for release and transformation. Coser concludes (1956, 157): "What threatens the equilibrium of such a structure is not conflict as such, but the rigidity itself which permits hostilities to accumulate and to be channelled along one major line of cleavage once they break out in conflict."

Wirth's (1949) and Coser's (1956) discussions on tension and conflict bring to the fore the importance of differentiating between tensions that result in creative and destructive conflicts, which in some circumstances are a matter of perspective of the agencies involved, but in another, a matter regarding the substantive content of disagreement and the conditions surrounding the disagreement.[2] How do we know when tensions are bringing about potentially creative conflicts and thus, should be encouraged? Do creative conflicts stay creative throughout or can they evolve to become destructive, and vice versa? To address these questions in the context of social relations in diversity, I find Baxter and Montgomery's (1996) discussion of contradictions (which I interpret here as tensions) in interpersonal relationship formation helpful to discern the ambivalence that is inherent in the concept of tension.

In *Relating: Dialogues and Dialectics* (1996), Baxter and Montgomery identified two kinds of contradictions. *Dualistic contradictions or tensions* are produced when opposites encounter each other. Dualistic opposites are "conceived as more or less static and isolated phenomena that coexist in parallel but whose dynamic interaction is ignored," such as in an "either/or" configuration (Baxter and Montgomery 1996, 10). Dualistic tensions are thus produced when differences are discursively framed as dichotomous rather than being engaged in a more nuanced and interactive negotiation. In contrast, *dialectical* or, more accurately, *dialogical tensions* are produced in an open-ended interaction between different but interdependent parts. Instead of dialectics that frame the interplay of oppositions as moving towards synthesis, dialogics per Bakhtin[3] emphasizes the indeterminacy and multivocality of social life as "both/ and" oppositions that are constantly interacting. Dialogism focuses on the process of dialogues to increase awareness and mutual understanding of the parties involved without requiring that a synthesis or common understanding be reached.[4] In personal relationships, dialogics are akin to relational dialectics. According to Baxter and Montgomery (1996, 9),

> Individual autonomy and relational connection are unified oppositions ... Connection with others is necessary in the construction of a person's identity as an autonomous individual, just as relational connection is predicated on the continuing existence of the parties' unique identities.

I will discuss the kinds of tensions experienced by the residents in the three multi-ethnic and multinational locales of Los Angeles, borrowing from Baxter and Montgomery's (1996) framework of relational tensions. As Wirth (1949) pointed out, social change triggers tension. Similarly, in twenty-first-century Los Angeles, tensions in demographically diverse locales are complex outcomes of sociocultural differences interacting with other processes of urban change, such as gentrification and the valorization of global capital, to produce heightened experiences of inequality and inequity between groups (Sassen 1996, Holston and Appadurai 1996). These experiences can deepen existing tensions and open new fractures and tensions in the social life of diverse locales.

Competing Values in San Marino

Over the course of four months of interviewing residents in San Marino from different ethnicities and nationalities, I heard residents mention a recurrent theme of social tension that had emerged from diverging cultural values and practices between neighbours. In my conversation with Dominique Fisher, a measured and lovely elderly woman who had lived in San Marino since the 1970s, she vividly described the social changes that San Marino had undergone. She felt that the change in the social life of San Marino was not only an outcome of demographic shifts but also of a profound change in the community's cultural values and practices. In this short excerpt from the conversation with Dominique, she spoke with candour about her personal experiences of living in a diversifying neighbourhood where social life was somewhat fraying from within. In this excerpt, Dominique also expressed a poignant disappointment with neighbourly relations, and like many of the residents whom I talked to, she regarded cultural difference that accompanies demographic diversification as the cause of divergence.

> Everywhere is safe, everywhere is home, everywhere is a place I feel comfortable. Now I am not sure if you talk to every Caucasian, they would say the same. I would hope they say the same thing. But I am sure they wouldn't. I think that people don't embrace change and don't embrace a multi-ethnic mix and a multicultural approach to life. They want it the way it was, and they are out of touch with reality. It makes me very sad. I think they miss out on a lot. I think that any of the people that I've met and talked to that are to some degree welcoming and comfortable but there are some, you know ...
>
> We had a house to the north of us a number of years ago. A Chinese family bought it and they totally remodelled the house. They finally moved in, which seemed like forever to me. I went over and said "Hi" as I would to anybody. They were there probably a few weeks ... and then they moved out. This was probably two years ago, and they never came back. They own the house. They'd done nothing with it. I mean, I think that is a cultural thing. And it's sad but I don't get it. They have lights that go on automatically, the gardener comes ... But since that time, nobody lives there. I mean they never came and said "We are leaving" or anything like that. And I think it's cultural ...
>
> The house across the street from us, I know they are inhabited by the helicopter kids. They live there, and they go to school. We've never seen them. It's sad. There is no interaction in the neighbourhood, the way it used to be. And I think it is a cultural thing. The houses are prepped up, and they are beautiful, but you never see a soul. Okay, you see the car, but you never see anybody ... I don't know it is very strange. To me, it is a cultural thing, and you don't like to do anything against the cultural mores.

In the conversation excerpt with Dominique above, the word "cultural" is repeated multiple times. Cultural difference is cited to explain the divergence of values within San Marino. At a deeper level however, this excerpt reveals a frustration at the insufficiency of multicultural relativism to address the social tension arising from difference within the community. Unpacking the interviews, I discovered that the "cultural" explanation that had been identified by residents as the source of social tensions, converged on two issues – property and parenting – that were not just "a cultural thing" but pertained to the values of economic, social, and spatial significance.

Value of Property: Use and Exchange Value

One of the prominent contentious issues in San Marino concerns the value attributed to property by its residents. In fact, when residents were asked if San Marino had any kind of social or cultural concentrations given its diversity, the responses of most residents were focused less on ethnic mix (save for a handful who marked out small White and Asian concentrations), but more on socio-economic differentiation – where the concentrations of wealthier and less wealthy residents live. As a premium residential address in a global city, San Marino's prestigious real estate attracts worldwide investment and thus, the locale is tightly connected to the flows of global capital. As a result, ongoing construction of new homes and renovation of old homes in San Marino has intensified the social tension between those who invest in San Marino as a node of global capital accumulation that otherwise would be footloose, and those who are investing in building up the local social capital in San Marino through stability and rootedness.

Over the years, a considerable number of habitable but modest-size houses have been demolished and completely rebuilt into large mansions by new Chinese investors. In many instances, these new mansions are unoccupied and regarded as a form of capital investments by their owners. As a result, these new homeowners are profiled by San Marino residents as foreign investors who are interested in building big houses (rather than yard space) to increase the resale (i.e., exchange) value of their homes. There is no intention on the part of these investors to live in San Marino. As a result, the mansionizing of San Marino has met with much disapproval among San Marino's long-term and older residents who have chosen San Marino for its use value of lower-density and greener surroundings. As set out by its founders, San Marino is envisioned to be a garden city valued for its greenery, stability, and preservation of its landscape.

According to the mission of the City Council of San Marino[5]:

> San Marino was formed to protect your personal rights and to control the growth and activities of the City in such a way that each individual resident will be guaranteed a pleasant place in which to live with a minimum of nuisance, with assurance that his property values will be protected by stringent zoning regulations. The principles upon which the first City Council established San Marino in 1913 have been followed through the ensuing 100 years. The founders of this city wished it to be uniquely residential – single-family homes on large lots surrounded by beautiful gardens, with wide streets and well-maintained parkways. There were to be

no manufacturing districts, heavy business areas or any apartment houses or duplexes. To maintain these standards, the City Council has continued to pass and enforce strict zoning regulations.

The social tension experienced is on multiple levels. The tension is between the transience and permanence of San Marino's different homeowners, as well as a tension between the different perceptions of homeowners about the value of their property, i.e., those who seek out exchange value versus those who desire use value of property. However, as Logan and Molotch (1987, 18–20) incisively pointed out, the contradiction is not so stark and dualistic:

> The stakes involved in the relationship to place can be high, reflecting all manner of material, spiritual, and psychological connections to land and buildings … Contrary to much academic debate on the subject, we hold that the material use of place cannot be separated from psychological use … Homeownership gives some residents exchange value interests along with use value goals.[6]

In other words, every property has a mix of exchange and use values for its homeowners. This means that each homeowner experiences a tension between exchange and use value of their property that is more dialectical in nature than dualistic. This inherent tension in values is projected outwards onto a neighbourhood level as well because the exchange and use value of a property is shaped and determined by the residential amenity of the locale.

During my interviews with the city planners, they similarly offered a nuanced picture of the contest between use and exchange value among homeowners. These planners explained that many Chinese homeowners had proposed new guesthouses or bigger houses because of the practice of living with extended family under one roof. However, recognizing that values on property were contested and contentious, and that the value of real estate in San Marino was intrinsically bound to the priority placed on the preservation of residential amenity and symbolic exclusiveness that the city provided, the city planners reaffirmed the need to ensure that construction of new houses was highly regulated and assimilated to the Euro-American aesthetics of what San Marino was known and cherished for. The city planners emphasized that San Marino must never go the way of Arcadia. Northeast of San Marino, it had transformed from a city with modest-size homes into an example par excellence of mansionization fuelled by Asian immigrants' pursuit of the American Dream. Lung-Amam (2017) wrote about a

Figure 4.1. (*left*) A modest 3,700-square-foot San Marino house up for sale for US$3.6 million, and (*right*) an 11,000-square-foot Arcadia mansion for sale for US$6.4 million in 2018.

From Moveto.com and californiarich.com (websites no longer available).

similar phenomenon called McMansions in Fremont, California, where high-income Asian immigrants had settled and built large homes to accommodate bigger families and reflect their aspirations. Akin to what happened in San Marino, these values were at odds with the local design standards and aesthetics expectations that were inclined towards those of White residents. Arcadia is, in short, like Fremont, the antithesis of what San Marino is envisioned to stand for. See figure 4.1.

In San Marino, the different set of use values that homeowners have about property ownership and neighbourhood amenity has become a source of consternation and social tension. Different customs, practices, and aesthetic expectations of what a house in a place like San Marino should look or feel like have pitted homeowners against each other. Lydia Li, a Chinese American resident in San Marino since the 1990s, attributed this outcome to the clash of "different sensibilities" for beauty, design, and maintenance of houses and gardens between homeowners. The conflict in tastes and care of property is further exacerbated by the financial ability of homeowners to implement their aesthetic and habitual preferences quickly and in a grandiose manner. For example, some Chinese homeowners in San Marino practice Chinese geomancy or *fengshui* in the design of their property. *Fengshui* propounds the intimate relationship between the positioning of one's property and one's prosperity. Following a master geomancer's *fengshui* advice, homeowners have removed trees in their front and backyards that can potentially block the flow of prosperity from entering their homes. Given the pride of San Marino as a city that is "uniquely residential – single-family homes on large lots surrounded by beautiful gardens, with wide streets and well-maintained parkways" (quoting the mission statement of the City Council of San Marino), the wilful tree removal has been a constant tension source between neighbours. It pits the values of Asian immigrant homeowners who believe that trees block the flow of prosperity against the values of other residents who value the garden city's large trees for the well-being they provide to daily life in a sprawling metropolis.

According to Sandy Cheng, an Asian American resident who lived through the early and tumultuous transition of San Marino from a predominantly White community to a more diverse locale in the early 1990s, the garden city versus prosperity geomancy controversy had been a hot button issue from the beginning between the Chinese and non-Chinese homeowners. The Chinese Club of San Marino took the lead to intervene diplomatically by conducting information sessions with new Chinese homeowners to educate them about the social and cultural value of trees held by others. In addition, clearer municipal guidelines on the removal and pruning of trees have helped to reduce the occasions for friction. Nevertheless, tree removal has remained an ongoing source of social tension in San Marino that divides neighbours into dualistic camps. This has resulted in a tree preservation ordinance on tree removal and pruning introduced by the city, among the many rules that have kept San Marino tidy.[7] In August 2011, an amendment was added to the tree preservation ordinance to impose new rules and penalties on the removal of trees in the backyard. Once again, in May 2018, the topic of tree preservation and removal was broached by the city council and a decision was made to prepare a new ordinance "in order to better protect and preserve the City's urban forest" (Section 1D) that was perceived to give San Marino its community identity.[8] The new ordinance on tree preservation and amendments were put in place in January 2019.

Consequentially, a consistent portion of the workload of the San Marino's planning office is in mediating divergent property sensibilities between its residents. The Planning Commission, which is viewed as the "first line of defense in protecting the physical character of the City," regularly meets once a month, while the Design Review Committee meets twice a month to "carefully review new and remodelled structures to ensure architectural compatibility with the neighbourhood" (City of San Marino, n.d., 5). During the meetings, residents are given an opportunity to discuss their preliminary proposals and to openly negotiate their disagreements with committee members who are resident experts in construction and design. The discussion at these meetings is taken very seriously by residents as the physical look and atmosphere of their neighbourhood manifests the economic, social, and cultural values of homeowners. Seen from another perspective, these regular and frequent committee meetings provide the tangible possibility to circumvent divisive, dualistic tension and transform it into more productive, dialogical tension through community-wide discussion. These meetings are part of the process of what Nicolaides and Zarsadiaz (2017, 344) termed as "design assimilation" of property and the built environment in their study of San Marino vis-à-vis other Asian American suburbs in the San Gabriel Valley, where the "preemptive campaign by longtime Anglo leaders to codify Euro-American design precepts, and the Chinese embrace of those values" are operationalized and reified.

In other words, the conversion of dualistic tension to dialogical tension over property takes more than community-wide discussion alone but an alignment of use and exchange values over time. The Chinese residents in San Marino have come to welcome the growing use and exchange values of their properties that accompany the leaving of minimal ethnic traces in the landscape so that San Marino is spatially distinctive as an authentic and exclusive American locale in the sea of surrounding Asian ethnoburbs that are ubiquitous with McMansions, large Chinese script signage, and a lack of lush greenery.

Values on Parenting: "Helicopter Kids" and Education

San Marino is the image of the good American neighbourhood life for raising children. Residents prize the city for its safety, peace, and good education opportunities. However, the ethos of the good family life and rich community life in San Marino is eroding, according to its long-time residents, the majority of whom are elderly White Americans who have lived in San Marino since the 1970s. During the interviews, the elderly residents spoke about how San Marino was no longer the way it used to be. They reminisced about the past: children walked or rode their bikes to school; neighbours chatted and shared their lives in their front yards while their children played together; block parties were frequent; and children played with each other in the streets. San Marino, from their point of view, had thin or non-existent neighbourly relations, and it had become glitzy like Beverly Hills. Instead of riding bikes to school, kids were being chauffeured by parents, often in spanking new luxury cars. This new image of San Marino becoming like Beverly Hills was unwelcome as it clashed with what San Marino stood for – down-to-earth, wholesome, and conservative.

Recognized as one of the best school districts in the United States, San Marino continues to attract many families who prize education highly, even those from out-of-state and from foreign countries. In the 1990s, "helicopter kids" or "parachute kids" from East Asia started to arrive in San Marino, according to the long-time residents whom I interviewed. "Helicopter kids" in San Marino is synonymous with well-to-do immigrant children from Taiwan, Hong Kong, or China who are "dropped off" in the United States for a better education and life opportunities, while their parents continue working halfway across the globe. These children are sometimes left alone, with a single parent, or under the guardianship of older siblings, relatives, or nannies. I was told that absentee parents tended to lavish their children with expensive cars and that the "helicopter kids" were reputed to have poor behaviour. A few White American participants also observed (with disapproval) that Asian children often stayed alone in the neighbourhood library until its closing time at 9:00 p.m., when their parents returned home to San Marino from work.

The lack of parental supervision for Chinese immigrant children did not sit well with the American parents who chose San Marino for its family-focused community identity that presumably also highly values proper parenting. In addition, the low participation rate of Chinese immigrant parents in the parent-teacher associations (PTAs), despite the growing immigrant students' enrolment numbers, has become a point of tension. In San Marino, active involvement in PTAs is a tacit civic responsibility expected of all committed residents with children in schools. Good parenting and civic responsibility are cherished community values. The increasing incidences of breaking social norms, not meeting the common expectation for participation, and thus undermining the cherished values of group have riled up dualistic tension and further entrenched the group divisions between residents.

Apart from the diverging worldviews on parenting, the education priorities and socialization preferences of Chinese immigrant, White American, and Asian American parents are also different. Chinese immigrant parents tend to focus purely on academics, differing from White American and Asian American parents, who tend to value sports participation, which they believe contributes an important part of a well-rounded education.[9] The focus on academics has led to Chinese immigrant children participating only in after-school tutoring, and not in recreation activities where they can interact with other non-immigrant students. The emphasis on education as purely academic achievement is a familiar phenomenon also discussed by Lee and Zhou (2015) in their research about what is cultural about the Asian American achievement. They found that among Chinese and Vietnamese immi-grant parents, they employ a set of common strategies to support the academic success of their children including buying homes in good school districts, pursuing honours and advanced placement classes, enrolling their children in intensive prepping and after-school tutoring, and finally, tapping into ethnic resources and strategies to gather infor-mation about neighbourhood quality, school rankings, and additional education opportunities. For Chinese immigrant parents in San Marino, giving their children the best academic opportunities is a form of good and responsible parenting.

However, these thin socialization patterns between immigrant and non-immigrant children have several consequences for the social life in San Marino: first, immigrant children are being prepped by after-school tutoring to be way ahead of their cohorts in the public schools. The knowledge disparity between students creates a tense and competitive atmosphere in the classroom. Some Asian American and White Ameri-can parents told me that teachers in the public schools were finding it increasingly difficult to engage and fruitfully teach a class of students in which some students had already acquired knowledge in advance due to after-school tutoring, and others had not. These parents also spoke about having the intention of taking their children out of the public schools and enrolling them in private schools because the education atmosphere is no longer productive to learning. In addition, they felt that with an increasing proportion of Chinese students in class, the pub-lic schools had become less culturally diverse. In fact, the decreasing diversity had become a concern for second-generation Asian American parents of Chinese descent, who felt that San Marino was becoming too ethnically Chinese and not the diverse place they wanted to raise their children in.[10]

Second, the divergence between the social worlds of Chinese immi-grant children and other children is further exacerbated by their choice of recreation, reflecting the divergent routine patterns of adult residents in the locale as illustrated in figure 3.4. Isabelle Anson, whose children attended public schools, noted that Chinese children tended to participate in orchestra, swimming, and tennis while White American children participated in football and cheerleading. In music, Chinese children gravitated towards playing violin and piano only. In addition, Chinese children took Chinese language classes instead of Spanish, which was popular among other children. Due to the different recreation activities in the schools, the opportunities for intermingling among the immigrant Chinese, Asian American, and White American children was minimized. Furthermore, during the weekend, the Chinese School classes were in full swing, meaning that many Chinese children (immigrant and non-immigrant) who attended these classes could not participate in the practices and games for the local Little League Baseball that were held during the same time.

The conditions for sustaining better inter-ethnic relations is made even more complex because the values of property, education, and parenting are entangled in San Marino, akin to many other locales where school districts draw on local property tax to finance their programs. As Jennifer Meier, a real estate agent and a long-time resident, explained to me, the property values in San Marino had increased over the years because the Chinese immigrant residents had kept the education standards of public schools in San Marino extremely high due to their emphasis on academic success. This had kept the property values in San Marino higher than those of its surrounding cities as San Marino had acquired a reputation of not only looking good as a safe and upper-class White American neighbourhood but also having high public education status. The intricate dialectical entanglement between people and property in this diverse locale creates a set of conditions that can foment protracted social tensions, which tend toward a dualistic bent.

Ethnic Turfs in Central Long Beach

Gang-related street shooting and robbery have plagued Central Long Beach for many decades. At the time of the interviews in 2011, residents spoke about how they consistently feared for their personal safety during daily activities of shopping, walking to school, and coming to work in Central Long Beach. Personal routines and community meetings within the neighbourhood were planned to end before nightfall due to the heightened awareness about the lack of public safety in the area. Anxiety, unease, scepticism, and tension filled the air in Central Long Beach.

On a late afternoon in November 2011, I chatted with Eric Alvarez, a community organizer, about social life in Central Long Beach. It was a memorable conversation in many ways. The November air was crisp and the setting sun was casting its warm, orange Californian glow on the grassy turf behind us. Central Long Beach looked surreally peaceful, and the atmosphere was relaxed, although twilight was approaching quickly. Eric jolted me out of that tranquil mood, for he wasted no time in getting to the struggles of social life in Central Long Beach. He told me that my question about social life was irrelevant without understanding the gang territories and their effects on the daily lives of residents. Eric reached over, picked up the Google Map of the locale in my hand and started sketching the geography of the gang territories. As he sketched, he introduced the active gangs to me.

> The area west of the flood control, that was the area called Westside Lon-
> gos. That's another gang area. But that's how Hispanic folks interpret
> neighbourhoods. They don't interpret it as like you live in the east side
> of Long Beach or west side Long Beach. You live in the Rollin 20s Crips
> neighbourhood. That's how they communicate it. That is how they inter-
> pret it. The kids around it.

A few weeks after speaking to Eric, I managed to get an interview with a patrolling police officer for Central Long Beach. To my surprise, the officer produced a map of gang territories that was similar to Eric's sketch map. Since then, I have always suspected that Eric is part of the gang intervention team active in the neighbourhood. My conversation with the police officer further ascertained the fact that gang territories significantly shape the form, geography, and possibilities of social life in Central Long Beach. He explained that the gangs in the area were 90 per cent ethnically homogeneous, so territorial gang activities further deepened the consciousness of ethnic territories. For example, Martin Luther King Jr. Avenue was notorious as a dangerous street where Black American gang territory overlapped with different Latino gang territories, becoming the site of violent clashes. The street was identified by several residents during the cognitive mapping interview as a corridor of crime and violence, and residents avoided using it if possible, especially after sunset.

The situation in Central Long Beach reminded me of a study of a multi-ethnic neighbourhood named Dover Square by Sally Engle Merry in her book *Urban Danger: Life in a Neighbourhood of Strangers* (1981). Merry (1981, 143) described the palpable sense of danger experienced by the residents of Dover Square, which shaped their mental spaces of the neighbourhood and affected their daily spatial practices.

> It is not simply the risk of crime that Dover Square residents find danger-
> ous, but the chance of random, vicious, unwarranted attacks by a stran-
> ger who belongs to a hostile group. Dangerous experiences include far
> more than crime. Insults, mockery, racial slurs, harassment, and flirtatious
> sexual comments that assault a person's sense of order, propriety, and self-
> respect awaken feelings of danger even though they contain no threat of
> actual physical violence.

The young adults who had grown up in the area developed an acute awareness of territoriality that helped them navigate the streets to avoid dangerous experiences. Like trained spatial ethnographers, they had a clear mental map of where the boundaries of ethnic enclaves overlapped and who belonged where. These young adult participants were able to point out which street blocks had high Black American occupancy, the locations where the Mexicans were mostly concentrated, and which apartment blocks were Cambodian vertical enclaves. For the youths, developing this keen street intelligence of ethnic territories was critical to surviving the rough neighbourhood, such as avoiding recruitment or attacks by gangs. Entering the wrong ethnic space at the wrong time could mean life and death for these youths. Street wisdom was essential to ensure personal safety (Anderson 1990).

Central Long Beach is a social space where gang territories overlap each other, and adding to that complexity, these gang territories interweave with pockets of ethnic enclaves so one's neighbours could be members of a gang, or for that matter, one's street could be an ethnic gang's territory even if the residents were not. Figure 4.2 shows a collective mental map of the gang territories layered over different spatial concentrations of ethnic groups and spaces of danger as identified by the residents according to their lived experiences and everyday spatial practices. The social space of the locale is conceived as a patchwork of ethnic territories rather than a neighbourhood unit with clear and singular boundaries. An analysis of these geographies reveals that where the gang territories overlap and collide, these spaces coincide with the danger spots that residents had identified and avoided.[11]

Twentieth Street in Central Long Beach is an example of this. Los Angeles's various infamous Black American city gangs, such as the Rollin 20s Crips, and the Insane Crips, as well as the Latino gang Eighteenth Street, have been expanding southward into the Latino East Side Longos gang territories, where Twentieth Street is located. Turf wars between these gangs are frequent. They range from racial slurs to violent fighting at night. Ben Rodriguez, a big Mexican American in his early twenties, had been living on Twentieth Street for ten years. Although he had respectful and friendly next-door neighbours, his experience along the street was a complete contrast. He told me that he would avoid walking along Twentieth Street alone at night due to bad experiences of being thrown dirty looks and verbally abused with racial slurs by people hanging out on the street. The danger of Twentieth Street also emerged in another interview. Calvin Jenkins, a frank and measured elderly Black American man who had just retired and temporarily moved to Central Long Beach, witnessed fights while driving along Twentieth Street one night. He recounted the event vividly to me:

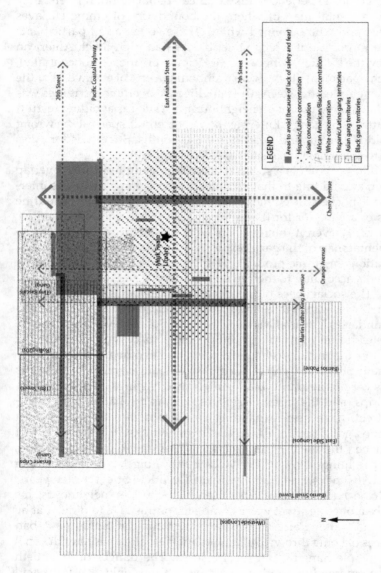

Figure 4.2. Map showing Central Long Beach as a multi-ethnic area with overlapping gang territories and danger zones. Map is not to scale.

Prepared by author.

It is rough. One night, I was driving at 9:00 p.m. at night. One guy was running away from another two guys. They were carrying sticks and rocks ... It shows my biases. They appear to be poor and struggling and young. They don't appear to have jobs looking at the way they dress. It is not fair to do it, but that's the way we are.

The continuing presence of ethnic gangs over the last fifty years in Central Long Beach and their continual violent activities to assert territorial control have fragmented the social space of the area into ethnic enclaves and injected distrust and anxiety into inter-ethnic relations. These feelings of unease and tension in the presence of the other further reinforces stereotypes and minimizes spontaneous inter-ethnic contact, resulting in a dualistic framing of relations between "us" and "them." These tensions reverberate through the locale to impose a spatial and social order upon the residents, which limits their daily geographical mobility and the opportunities for dialogical intercultural learning between neighbours.

At the time of fieldwork, annual community events such as the Martin Luther King Jr. Parade organized by the city hall promoted the sociocultural diversity of the City of Long Beach, and the programs at local recreation centres provided opportunities for reconciliatory dialogic tensions to form among the youths from different social and cultural groups. However, the production of dialogical tensions is limited to these occasions and to those willing to participate. During my visits to a local recreation centre popular with the elderly in the locale, the preference to "hunker down" (Putnam 2007) in one's ethnic group was evident, whether in dance classes, playing pool, or eating lunch. The lack of a common language and preference for familiarity curtailed interaction between members of different groups. Although the neighbourhood recreation centre was a safe space, the dualistic tensions formed between ethnic groups as a result of decades of ethnic gang violence in the open streets outside had instilled a habitual scepticism and indifference to inter-ethnic relations. In this regard, residents in Central Long Beach live "parallel lives" (Cantle 2005) because the strong dualistic tensions within the locale exert a centrifugal force upon inter-ethnic relations.

Profiling in Polarities: Mid-Wilshire

Mid-Wilshire, a neighbourhood close to the glamour of Hollywood and Beverly Hills with a history as a major site of the 1992 Los Angeles civil unrest, is a locale with multiple lines of tension arising from the myriad of socio-economic and sociocultural differences found within its borders. One of the salient inter-ethnic tensions in Mid-Wilshire experienced by participants is the atmosphere of public unease with the presence of Black Americans in the area. As Courtney Bateman, a long-time retail proprietor in western Mid-Wilshire explained,

> I think that there are very few African American people here … In Larchmont recently in one of the homes, there was a home burglary. Actually I don't think they got in but the cameras captured the individuals. The two individuals were African American men in their twenties, and then there was some crime on the Boulevard [referring to upscale Larchmont Boulevard] recently, and there were photos of an African American man … and so I say amongst owners of businesses on the Boulevard, if they see an African American man in their twenties, they are likely to look him over very carefully and be very suspicious because of the recent crimes and because of our general society's stereotypes about African American men in their twenties … The diversification of this area is African American challenged.

During the interview with Sergeant Caleb Torres from the Los Angeles Police Department (LAPD), he explained that many business owners in Mid-Wilshire were fearful of the male Black American presence because of the repeated robberies committed by Black Americans in the area. In addition, the expansion of the "South LA Bloods" gang (a predominantly Black gang) northward into Mid-Wilshire had also increased the prostitution and drug activities in the area. Sergeant Torres felt that it was difficult for these stereotypes and prejudices to be removed in an area with a history of violence and ongoing inter-ethnic tensions.

> People don't have a reset button … If it did not happen to them, they can tell you about an incident that happened to somebody they knew and they carry that with them … You hear those statements, "You know, they are all like that." "That's how they are." … I can't tell him or her that he or she is wrong. They are making it based on personal experiences they had or what they have seen … We can't push reset. Officers can't push reset either. People would love it if an officer or any person would judge a person based on what happened at that moment. But that is not possible because if it is baggage or experience, you carry that with you. You will be a fool not to. This experience can help us remain safe. You always have those things.

Sergeant Torres's poignant reflection made me realize that the shrinking Black American resident population in the area over the last few decades has additionally made Black Americans an unfamiliar and thus, suspicious presence. Charlie Brooks was one of the few Black American residents whom I interviewed. Charlie worked in the entertainment industry in nearby Hollywood, and he had lived in Mid-Wilshire for many years but in different locations. At the time of research, Charlie had recently moved into an upscale rental apartment in western Mid-Wilshire. Over the course of our conversation, he spoke about his experiences of racial discrimination in the area, especially those that happened in the hip, expensive restaurants and shops along Larchmont Boulevard – an area usually patronized by White American and Korean clientele. Charlie recounted the two times when he was refused service by restaurants at Larchmont Village and thus, he rarely visited the shops along the Boulevard even though he lived within walking distance. He contrasted these negative experiences with the positive experiences in the American Korean eateries in nearby Koreatown, where he was a regular customer. He related to me with pride about how he felt welcomed at the restaurants, and he even befriended a Korean chef at a Korean pizza joint who would welcome him like a friend whenever he visited there.

In Mid-Wilshire, intra-ethnic antagonism among different nationalities of migrant Hispanics further complicates the formation of social life. Guatemalans, Salvadorans, and Hondurans in the area consciously differentiated themselves from the Mexicans. I learned this from an El Salvadoran 1.5-generation immigrant who had grown up in Mid-Wilshire. He explained to me that Central American immigrants were bullied and badly treated in Mexico during their migration journey northward into California, and their bad experiences had made many resentful of the Mexicans. Thus, the Central American residents in Mid-Wilshire actively rejected being associated with the term "Latino," preferring the term "Hispanic" to distance themselves from being mistaken as a Mexican. Even among the second-generation participants who were born and raised in Mid-Wilshire, the preference to use "Hispanic" instead of "Latino" in their self-description was clear. When asked how they identified themselves, they named their parents' national origins in Central America.

These dualistic tensions between different Hispanic nationalities are further entrenched by violent gang rivalry between Central American and Mexican gangs. I got to ride along with Sergeant Torres in a LAPD patrol car as he went on his patrol; during the ride he pointed out many incidences of gang tagging on the walls of apartment buildings and at street corners. It was particularly evident in the border zone between two gang territories, and Sergeant Torres interpreted the code words of gang aggression in the graffiti for me. Turf wars between the Eighteenth Street gang (a predominantly Mexican gang) and the rival MS Gang (made up of members from El Salvador and Honduras) are frequent in the area. The prevalence of Hispanic gang activities such as drug dealing and gang tagging has, over time, caused the non-Hispanic residents, particularly those who have newly arrived, to equate "Latinos" and "Mexicans" with gang activities.

The behavior of Hispanic gangs had entrenched the impression among the non-Hispanic residents whom I interviewed, particularly the new arrivals, that "Latinos" and "Mexicans" meant gang activity. Residents like Michael So, a newly arrived international student from South Korea studying at a local community college and living in eastern Mid-Wilshire, felt fearful and anxious whenever he was near Mexicans. This anxiety had restricted his mobility in the area and significantly shaped his spatial practice and lived experience. Around the eastern Mid-Wilshire area that Michael lived in, there were enclaves of Hispanics and Koreans and frequent gang activities. In our interview, Michael described his daily experience in the city in heartfelt words:

> I don't know exactly but the other races, I have fear of them. When I drive to school, I see kind of Mexican place. I feel this is Mexican place, I just drive through. And when I see Korean thing, I feel this area is okay.

Shortly after speaking with Michael, I visited his part of the neighbourhood during my patrol ride with Sergeant Torres, and the visit helped me understand more fully Michael's anxiety about living in eastern Mid-Wilshire. During the ride, a sudden call for Sergeant Torres to assist with suspected criminal activity turned the calm patrol ride into a high-speed chase. We sped through the crowded streets, at times against the traffic, with the full blast of the urgent siren above us. A man was suspected of possessing an illegal gun, and radio updates were constantly streaming in. He was on the run on foot and had tossed out packets of drugs. When Sergeant Torres and I finally arrived at the crime scene, a site very close to where Michael lived, a crowd of residents had already gathered around the traffic intersection. I was shocked to see a young man barely twenty years old being handcuffed and led away; multiple small packets of drugs were being gathered in a pile on the asphalt close by. In a nonchalant tone, Sergeant Torres informed me that the boy was a repeat youth offender whom the police recognized well.

Frequent criminal activities escalate an air of tension and felt anxiety in the neighbourhood. Perceivably, repeated occurrence of similar events like this palpably shape a resident's routine spatial practices to distance people from certain places and social groups, establishing and reinforcing stereotypes that in turn incur prejudice and tension between individuals from different social groups. This was akin to what Sergeant

Torres had shared earlier: in environments like this, it would be difficult to reset impressions and relations. In fact, learning where the danger spots are in the locale can help residents stay safe.

The intertwining of ethnic and national affinities, polarizing incomes, and lifestyle disparities make navigating diversity a challenging but also a fearful experience for Mid-Wilshire's residents. Residents, community organizers, and municipal officers described their experience in Mid-Wilshire as living in a landscape of anxiety and social tension. Being in the presence of many unfamiliar languages, cultures, and socio-economic polarities was discombobulating and uncomfortable, and it generated social anxiety for the poor, the middle-class, and the rich alike. An insight into the uneasy experience of negotiating coexistence in diversity came from Mark Adams, a Black American western Mid-Wilshire resident and an acquaintance of Charlie Brooks. Over a cup of coffee at Starbucks, Mark reflected on his experiences living in the locale, and underscored the entanglements of racism and income disparity that made everyday life in Mid-Wilshire a state of perpetual competition and dualistic tension. Pointing to the map of the locale that he drew on, Mark explained:

You would have it [conflict] in this area [*referring to poorer part of Mid-Wilshire*] because of the lack of education and insecurity. This area [*pointing to the western half of Mid-Wilshire*] at the spiritual level, they live in abundance. They understand that there is plenty for everybody. They have space, enough room, enough parks, enough water, and enough air. Here [*the eastern half of Mid-Wilshire*] people live in the opposite. They don't feel there is enough. So they are going to take. If this group going to move in here [*the western half of Mid-Wilshire*], what are you going to do? They [*eastern Mid-Wilshire*] live in fear, and they [*western Mid-Wilshire*] live in abundance ... When there is a feeling of lack or when your security is threatened, that is what causes racism. If someone moving in here [*western Mid-Wilshire*], it would cause major chaos and vice versa. It is socio-economic but when they physically show up, it becomes racial ... The White person will be given a benefit of a doubt and the possibility to get with the program.

As the composite map of the social territories drawn by the residents suggests in figure 4.3, the social space in diverse Mid-Wilshire is a geography of residential clustering and segregation along lines of ethnicity and income, rather than an evenly mixed landscape of cultures. In addition, the locations of "diversity" as experienced by residents are also delineated by them as places of higher income, i.e., western Mid-Wilshire. For example, the area of greatest sociocultural diversity is located between Western Avenue and Larchmont Boulevard, i.e., the area of western

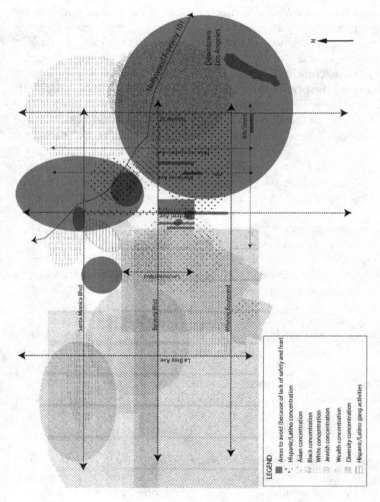

Figure 4.3. A composite map of Mid-Wilshire showing a fragmented landscape of ethnic concentrations and territories as conceived by its residents. Map is not to scale.

Prepared by author.

Mid-Wilshire. The diversity there is conceived by residents as over-lapping concentrations of wealth with a specific mix of ethnic groups, namely White, Asian, and Jewish. In eastern Mid-Wilshire, there are no areas of diversity. Instead, the geography is composed of a large Hispanic concentration, peppered with small pockets of Asians and Black Americans within. Significant swathes of eastern Mid-Wilshire are also considered dangerous and best to avoid by residents.

There were, however, some residents in Mid-Wilshire who felt that living amid multiple social and cultural differences was stimulating and exciting. For them, sociocultural differences did not create unproductive dualistic tensions; instead, these differences catalysed dialogic tensions that made them curious and motivated to engage in intercultural learning. These residents, mostly from western Mid-Wilshire, had actively chosen Mid-Wilshire for its sociocultural diversity. Their positive view of living amid diversity contrasted sharply with the residents in lower-income eastern Mid-Wilshire. Perceivably, with resources to choose the kinds of diversity to engage with and those to avoid, residents in western Mid-Wilshire could negotiate their differences from a place of safety and comfort.

In contrast, for the majority of the participants who lived in eastern Mid-Wilshire, the social space there was unstable, unpredictable, and filled with stressful dualistic tensions between groups. As a result, many of them chose to hunker down and not engage in forming inter-ethnic relations. The multiple intersections of palpable differences in eastern Mid-Wilshire created an intense atmosphere of the unknown, the unfamiliar, and the disparities (real and imagined) among ethnic and income groups, which generated undercurrents of fear and anxiety in residents to cross sociocultural boundaries. These forces destabilized the formation of durable intercultural relations and conviviality in Mid-Wilshire.

Concluding Thoughts about Tensions of Diversity

The findings from Los Angeles indicate the predominance of dualistic tensions across the different locales of diversity. These tensions exert divisive pressure on social life, making coexistence in diversity a constant work of negotiation. There is evidence of some dialogical tensions at work. When present, the tensions exert a counter force to hold relationships together, just like how centripetal force acts against the centrifugal force of a same object in motion.

Dualistic tensions in the three locales are associated with several broad processes at work in diversity. First, *social categorizing according to ethnicity and nationality* using pragmatic and general visual, verbal, and contextual cues are employed by residents to organize their lives in diversity. Social categories, according to Allport ([1954] 1979), Milgram (1970), and Lofland (1973), aid the human mind by giving order to complex social situations and help to overcome the stimulus overload that commonly occurs in contexts of heterogeneity. However, once formed, these prejudgment categories can become inert and static, turning into stereotypes. When that happens, differences between groups or individuals are simplified and become dichotomous or artificially enlarged so that we believe that group differences are larger than they really are. As Blaine (2013, 34) explains, quoting Rothbart, Evans, and Fulero (1978), the sheer effect of social categories is that differences between individuals are reduced to group traits because of our "tendency to confirm rather than disconfirm stereotypical thinking and expectations about other groups." Thus, our perception of the actual diversity of an environment is in fact dependent on how complex our categorization systems are. The implication of these mental social categories is that they can generate durable social boundaries between groups that either increases or reduces the contact opportunities between individuals and groups through the way boundaries are enforced, maintained, and negotiated (Barth [1969] 1998). This is observed in eastern Mid-Wilshire and Central Long Beach where the availability of ethnic bubbles and heightened distrust between groups have diminished the formation of dialogical tensions.

Second, the *practice of different and competing values* by residents from different social and cultural groups produce dualistic tensions that divide social life in the diverse locales. As Allport (1954 [1979], 25) explained, values are like fences and "are built primarily for the protection of what we cherish." Thus, values for individuals are safeguarded boundary markers that differentiate those inside from those outside. They are by nature dualistic, separating those who share them from those who do not. The findings in Los Angeles show that the residents in these three locales of diversity often attributed the differences in the behaviour of certain groups to cultural values as a causal factor. Blaine (2013, 38), quoting the work of Pettigrew (1979), explained that in situations where there are out-groups and in-groups, as in the case of diverse settings, behaviours of out-group individuals are often attributed to "inner, dispositional causes" rather than "situational, circumstantial causes." Salient issues arising from cultural value divergence include ethos of family life, parenting philosophy, education, and sensibilities toward property use, as demonstrated in San Marino in particular. In Central Long Beach and eastern Mid-Wilshire, residents spoke about the inward-looking, asocial, and unengaged nature of Asians as attributed to the Asian culture that valued family life above all else.

In contrast, where productive dialogical tensions are present, there is evidence of the *sharing of values*. Residents, particularly those in the higher income neighbourhoods of western Mid-Wilshire and San Marino, often spoke about the importance of having shared opinions, beliefs, and priorities in order for sustained interpersonal relations between individuals from different social and cultural groups to form. The residents in western Mid-Wilshire recognized that sharing simi-lar socio-economic and class statuses had enabled them to transcend the divisions created by ethnic categorization, on some level. In this way, the higher income residents in the presence of low-income resi-dents share a subculture within Mid-Wilshire. Subcultures, according to Fischer ([1976] 2005), proliferate in cities to enable new kinds of affili-ation between people as they transcend old identity moorings and ways of seeing. In San Marino where every household has an income above the county average, sharing similar socio-economic statuses is not quite enough for dialogical tensions to counter the fomenting of dualistic ten-sions. Memberships in interest-oriented social clubs, such as the PTA or Rotary Club, provide a necessary common platform that reminds residents of their shared resources and values. The club activities offer regular opportunities to break out of a binary narrative of framing rela-tions in the city as Chinese versus non-Chinese communities. However, despite the presence of these platforms, dialogical outcomes are limited, as the formation of subgroups – usually according to ethnic affiliation and immigration experience – continued over time to formalize a new set of social boundaries of inclusion and exclusion within the clubs.

As Los Angeles is a metropolis and immigrant gateway, its social environment is inherently complex and diverse. Each locale is chal-lenged with its unique bundle of conditions that shape how coexistence is negotiated. This chapter has outlined where the overlapping pat-terns of tensions of diversity in the locales lie. The prevalence of fleet-ing relations in Los Angeles, the reliance on "categoric knowing," and the entrenched presence of ethnic gangs interact with socio-economic inequality and racial prejudices to produce a social space prone to the formation of dualistic tensions.

Socioculturally diverse locales are environments embedded with "societal fault lines" of social, economic, and political differences that are intricate and insidious.[12] As these differences intersect, they pro-duce friction, instability, and tensions of both the dualistic and dialogi-cal kinds. These locales, as contact zones with multiple "societal fault lines" are dynamic spaces of social and cultural contestation, according to Pratt (1991), that require much finesse to harvest the potential of ten-sions for intercultural learning. In the next chapter, we will unpack the fault lines spatially by taking a close look at the construction of neigh-bourhood boundaries in the three locales and discussing its implica-tions for the fostering of local belonging.

5 Boundaries and Local Belongings

You are taught to stay away from certain people because you don't want a conflict. Instead, you tend to stay around your own, in your own little area, in your own cubicle.

Marteese Owens, Black American man in his thirties
Central Long Beach resident

Sociologist Saskia Sassen (1996) provocatively asked at the turn of the twenty-first century, "Whose city is it?" With growing population densities and demographic diversification, urban territoriality has become ever more salient in cities that actively seek to participate in the global circulation of labour and capital. We see new and contested claims to the use of urban space as the number of social and cultural groups increase. Embedded in this claim to space is the desire and need to belong. As anthropologists Gupta and Ferguson (1997) incisively noted, de-territorialization through global immigration fuels reterritorializing. Thus, the process of place-making, through an ongoing construction and maintenance of social, symbolic, and physical boundaries, has become a core aspect of negotiating belonging in cities undergoing social and cultural changes arising from globalization (Friedmann 2005).

My argument here is that the tensions in diverse locales cannot be explained solely as a social or cultural phenomenon but that these tensions are fuelled by the meanings we invest in our dwelling places. To quote geographer Yi-Fu Tuan ([1977] 2011, 138): "Place is a pause in movement. Animals, including human beings, pause at a locality because it satisfies certain biological needs. The pause makes it possible for a locality to become a center of felt value." Using the examples of the three locales, this chapter puts forward the notion that by understanding the residents' mental space of the locales via their conceived neighbourhood boundaries and their routine practices, we are better able to understand the socio-spatial significance of the need to belong in diversity, and thereafter, the possibilities for collective life.

In the immediate section to follow, I will discuss the different kinds of important boundaries in each locale of diversity – elective, circumscribed, and polarized. Analysing the cognitive maps through the filter of ethnicity, nativity status, and residency length, I found that there was no generalizable pattern of conceived boundaries along lines of ethnicity or nativity possibly due to the stratification of an already small sample size. Instead, the analysis revealed that boundary-making practice varies according to residency status, residency length, and income level when a comparative analysis is undertaken across locales.

Elective Boundaries

In my interviews with regular visitors to San Marino from neighbouring cities, whom I met at San Marino's beautiful Lacy Park and the spacious Crowell Library, they spoke about how they instinctively recognized the social exclusiveness of San Marino upon entering the city. They highlighted San Marino's distinctive atmosphere of calmness, quiet, and cleanliness that stood out among the surrounding cities that they lived in. One of the regular visitors I spoke to was Zack Shi. Zack, a polite and studious Chinese man in his late twenties, was studying at one of the large tables in the library when I met him. Zack is from China, and at the time of interview, he had just received a graduate degree in computer science from UCLA. Zack used to live on the Westside of Los Angeles to be close to the university but had just relocated to Monterey Park area (east of Downtown, near San Marino) to save on rent while looking for full-time employment. He was a regular user of the library who visited San Marino from a neighbouring city and liked the quiet he got at the library. Zack told me that he found out about this library a few months ago when he worked in an after-school tutoring centre in San Marino.

At the start of the interview, I asked Zack to locate the library on the map (with a star sticker). Then I asked Zack to draw the boundary of the neighbourhood from his point of view. Picking up a marker, Zack drew faint lines on the map that divided the neighbourhood into thirds, explaining the significance of these lines to me as he drew (see figure 5.1):

> You can see the boundary really clearly. When you go to Alhambra, Monterey Park, Temple City, you cannot even tell the difference between each other. When you go to San Marino from Alhambra or San Gabriel, you can see a really clear boundary [*drawing dotted lines*]. Like this part is really organized and have bigger buildings [*pointing to San Marino*]. And this other part the buildings are not [*pointing to Alhambra and San Gabriel*] … I

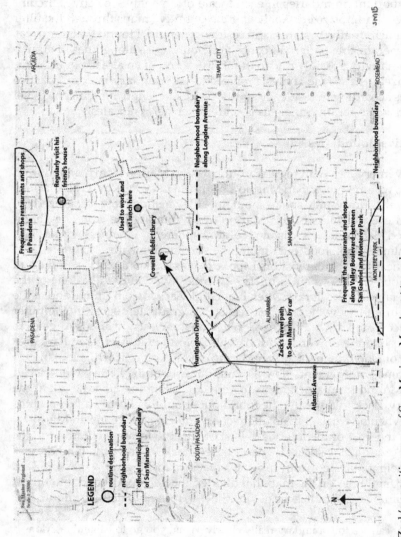

Figure 5.1. Zack's cognitive map of San Marino. Map is not to scale.

Labelling by author. San Marino municipal boundary added in by author to enhance map presentation.

don't know the boundary but when I drive the car past this area [*drawing a vertical line*], they have a stone there and they say that this part is San Marino. You can see it really clear – the difference between the two areas. Here [*in Alhambra*], you see the banners written by Chinese and the Chinese immigrants live here. There is a boundary [*drawing a line along Longden Drive, dotted*]. When you pass here you barely see a Chinese written banner.

For Zack, these lines distinguished San Marino visually and symbolically from its neighbouring cities. In his mind, San Marino belonged to the northern group of wealthier and socioculturally diverse cities that includes Pasadena, Monrovia, and Arcadia. These cities are recognized for their socially mixed communities that include White Americans, Latinos, and well-heeled immigrants from Taiwan, Hong Kong, Vietnam, and China. In comparison, cities south of San Marino, such as San Gabriel, Alhambra, Monterey Park, Rosemead, and Temple City (where Zack lived), are poorer enclaves that consist of immigrants from China as well as low-income Latinos. When we compare Zack's cognitive map in figure 5.1 to the dot density map of ethnic concentrations in the San Gabriel Valley in figure 3.1, we can see that Zack's description closely illustrates the census. San Marino's ethnic mix indeed resemble the northern cities of Pasadena and Monrovia more than the surrounding southern cities of Alhambra and Monterey Park, which have a high Asian population concentration.

When the maps drawn by visitors are compared against those drawn by residents, what stands out is the level of detail that residents gave to marking out the actual physical boundary of San Marino. Even though Zack visited San Marino almost daily, he was only able to perceive the broad differences between San Marino and the southern cities, and could just sense the approximate location of the actual municipal boundary between neighbourhoods. His mental image of difference between San Marino and its southern neighbours is stark, but on the ground the difference is visually gradual because the immediate residential streets surrounding San Marino have been designed to have houses and lawns that look very similar to those in San Marino.

In comparison, the residents' conceived spatial boundaries are almost identical to San Marino's official municipal boundary. The degree of congruency and precision of the residents' conceived neighbourhood boundaries – even along the northern boundary where the winding streets made it hard to discern the side of the road that belongs to San Marino – is indicative of the presence of a collective symbolic importance given to place-making among its residents. Compare figure 5.2 and figure 5.3 to see the difference between the collective boundary maps drawn by visitors and residents, respectively.

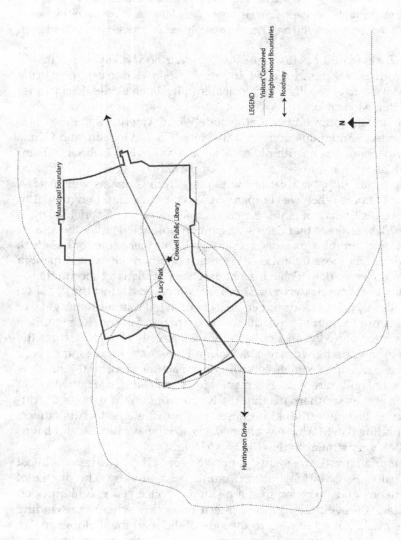

Figure 5.2. Map showing the collective conceived boundaries of regular visitors in San Marino. Map is not to scale.

Prepared by author.

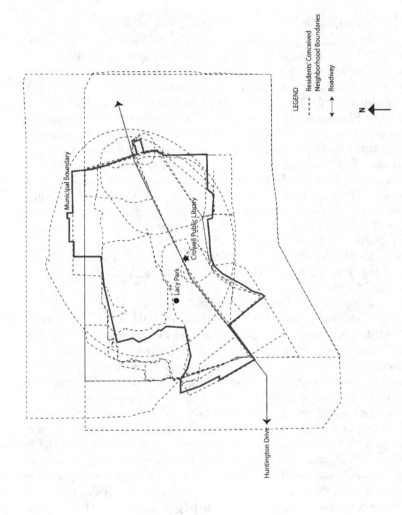

Figure 5.3. Map showing the collective conceived boundaries of San Marino residents. Map is not to scale.

Prepared by author.

My view is that the high level of similarity among the residents' conceived neighbourhood boundaries is not accidental but illustrates a process of deliberate place-making that asserts territoriality and a sense of elective belonging (Savage, Bagnall, and Longhurst 2005). Lefebvre's ([1974] 1991) conceptual framework of the triadic dialectical production of social space is particularly helpful here to unpack and explain the reinforcing dialectical interaction between the spatial practice (perceived space), mental (conceived space), and the symbolic (lived space) in place-making.

As a city within Los Angeles County, San Marino has an independent city council, police and fire department, clear jurisdiction, and exclusive zip codes. It is a place with clear physical and administrative boundaries, reinforced by official municipal documents and plans. More importantly, San Marino's exclusivity as a high-income city nested in a wider geography of cities with lower-income residents relies on its ongoing efforts to differentiate itself from the surrounding cities. San Marino does this through the construction of boundaries in ways that are perceivable for those living on the inside and those coming from the outside. Through the passing of tree ordinances and strict zoning regulations, such as those that limit the size of dwellings and the kinds of fences allowed on properties, San Marino's built environment is consciously crafted or conceived by the municipality so that the landscape of the city looks and feels different from its surroundings. This difference in the perceived space further enhances the symbolic exclusiveness of San Marino as a lived space.

San Marino is a prestigious address in Los Angeles with high property values and reputable public schools, and its residents elect to invest in it to become proud homeowners who can then claim a sense of belonging to a reputable city, whether or not they derive use values from their properties by living there. Having an address in a prestigious city is akin to owning a positional good by being affiliated with it, and symbolically becoming a part of it. They demonstrate a sense of "elective belonging" that Savage, Bagnall, and Longhurst (2005) refer to when people exercise their choice of neighbourhood for one that best articulates their identities and values. The individual's sense of symbolic, elective belonging to San Marino is intensified by municipal policies: for instance, residents may use Lacy Park for free during the weekends while non-residents are charged a fee, and residents enjoy preferential pricing for the annual Fourth of July celebration in the park. Through the conceived space of municipal policies, the city shapes the spatial practices of visitors and residents and practically maintains the symbolic feeling of "the elect." This sense of elective belonging, I think, is a major reason for the clarity of the mental neighbourhood boundaries of San Marino among its residents.

Circumscribed Boundaries

During the interviews, the residents in Central Long Beach used recurring concepts, such as "bounded territories" and "enclaves," to describe their experience of social life in the locale. The analysis of the cognitive maps in this locale reveals that residency length rather than the status of residence (as in the case of San Marino) shapes the conception of the boundaries most significantly. Long-time residents who had lived in the area from six to forty years, particularly those who had grown up in the locale, drew cognitive maps that demonstrated a high degree of specificity at the block level with details of major roads, local streets, and ethnic concentrations. In contrast, the cognitive maps of new residents who had lived in Central Long Beach between three months to five years had fewer details and indicated only the major roads that were familiar to them. While drawing these neighbourhood maps, new and old residents alike told me that their daily activities within the neighbourhood were circumscribed because of their heightened fear of street violence. Often, they conceived their neighbourhood space in terms of safe and familiar places.

Analysing the neighbourhood boundaries of different residents collectively, I found that in spite of the variations among maps, there was a strong convergence on a set of spatial boundaries around what residents conceived as the neighbourhood of Central Long Beach. Major street corridors, such as Cherry Avenue, Orange Avenue, Martin Luther King Jr. Avenue, Pacific Coastal Highway (PCH), East Anaheim Street, and Seventh Street formed the neighbourhood boundaries for most residents.

A cross-analysis with information of conceived territories showed that these conceived neighbourhood boundaries of Central Long Beach coincided with spaces of danger in the neighbourhood that the residents had identified. As pointed out in chapter 4 and as illustrated in figure 4.2, these spaces of danger were in fact areas where gang territories overlapped. During the interviews, the residents unanimously spoke about the vices on Pacific Coastal Highway (PCH). According to them, PCH was a strip infamous for drug dealing and prostitution, while Martin Luther King Jr. Avenue (between Seventh Street to PCH) was notorious for its gang activities, drugs, and shootings. Young Latino and Black American residents who wanted to stay clear of predatory Latino and Black gangs in the area had to avoid these streets at all costs. In addition, the area west of Cherry Avenue and north of Seventh Street, where most of the social and commercial activities for those living in Central Long Beach were located, was regarded as a no-go zone for those living outside the area. Kylie Brendon, a new White American resident who had recently moved into a house that was a couple of streets east of Cherry Avenue, told me that her constant advice to friends visiting the area was to stay on the east of Cherry Avenue. This was because Cherry Avenue marked the division that separated the better-off and safer neighbourhoods from the poorer and more dangerous ones in Central Long Beach. See figure 5.4 for the collective map of residents' conceived neighbourhood boundaries and the spaces of danger identified by them.

The matching geography of the conceived neighbourhood boundaries and danger zones underscores an important and troubling finding – these neighbourhood boundaries act like fortress walls and moats that physically and socially enclose its residents, separating them from those who do not live here. Unlike in San Marino where people from surrounding cities visit its park and library, there are few visitors in Central Long Beach. As a low-income neighbourhood within the City of Long Beach, its residents also face limited opportunities to afford the amenities in the surrounding neighbourhoods. I believe this is the reason some of the residents, in particular the younger ones, had unreservedly identified the neighbourhood as "a ghetto." These boundaries and spaces of danger conceived by the residents are not solely mental or abstract, but material in a real way. Often, they are memories of personal danger, of witnessing acts of aggression, as well as those of learning from friends, family, and police reports about street violence. These memories contour the mental maps of the residents that in turn actively shape their daily spatial practices in the neighbourhood. These conceived boundaries are multifaceted as they also function as an integrated spatial, social, and symbolic mechanism that moulds the practice of social interaction among neighbours, the possibilities for collective life, and the meaning of the neighbourhood as a lived space.

Polarized Boundaries

On a November afternoon at the Wilshire Branch Library, I met young second-generation Mexican American, Luciana Garcia. Luciana had lived in Mid-Wilshire for the last seventeen years and we spent some time chatting about her experiences growing up in Mid-Wilshire. She painted the details of the places in the area with great ease and enthusiasm. Luciana recounted her memories of living near the library during her childhood years about a decade ago and recalled vividly that the old apartment environs were quiet, "people dressed nice," and there were opportunities to speak English (see figure 5.5 for Luciana's map). In contrast, the area around her current apartment, located one mile (1.6 kilometres) away in eastern in Mid-Wilshire, had predominantly Hispanic residents. The area was always noisy, filled with big families, tired workers, and "people dress whatever." When asked if

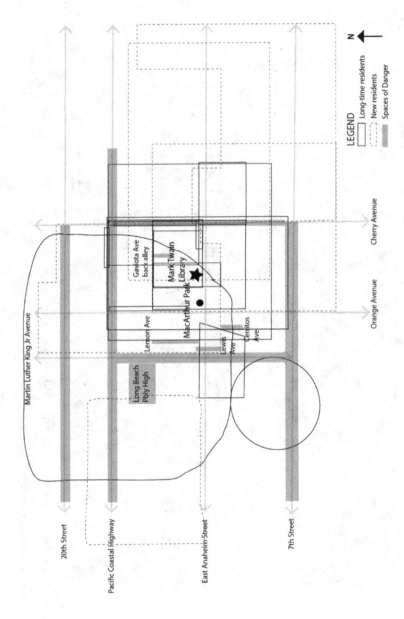

Figure 5.4. Map showing the collective conceived neighborhood boundaries of Central Long Beach residents. Map is not to scale. Prepared by author.

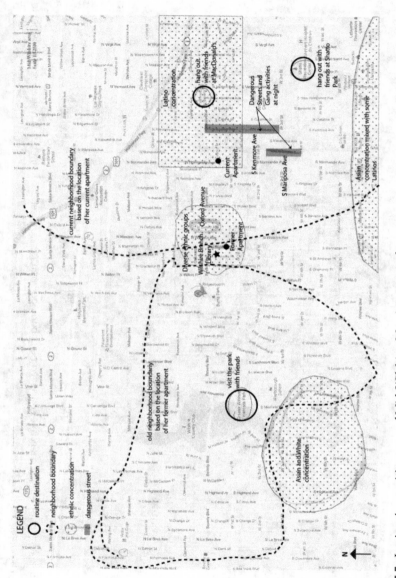

Figure 5.5. Luciana's cognitive map of Mid-Wilshire. Map is not to scale.
Labeling and map graphics enhanced by author.

she thought the Mid-Wilshire area as a whole was diverse, Luciana responded,

> From this neighbourhood [area around the Wilshire Branch Library], I believe that it is diverse somewhat because right here it is filled with the White population. But when you head to Beverly [from Beverly Blvd./Normandie Ave. towards Downtown Los Angeles], you are heading to the Latino community. If you head down from here to Beverly Center [towards the west], you feel like it is diverse because you see mostly Asian and White communities. It is weird because you are in a community you feel like you don't belong to. But when you are there [pointing to the Beverly Blvd./Normandie Ave.], people are more friendly, you talk to more people. You know it. You are used to your own colour.

Luciana is not alone in perceiving, experiencing, and conceiving the socio-spatial divisions in the diversity of Mid-Wilshire that she had described. Other residents, from both eastern and western Mid-Wilshire, talked about similar experiences of geographical and social distances between the two halves. For example, residents living in eastern Mid-Wilshire (east of Western Avenue) conceived their neighbourhoods as geographically smaller units that ranged from three to seven blocks in radius. This is likely due to the low levels of car ownership among the residents in eastern Mid-Wilshire, which limited their mobility, and the high population density in the area, which can reduce one's sense of familiar space.[1] In comparison, most of the residents in western Mid-Wilshire (west of Western Avenue) used private automobiles to get around, and they conceived their neighbourhood to be more extensive. In addition, the collective analysis of the conceived neighbourhood boundaries by residents of both eastern and western Mid-Wilshire shows that Western Avenue acts as a wall that divides Mid-Wilshire in two halves. See figures 5.6 (eastern) and 5.7 (western).

A cross-analysis of the routine and spatial practice maps of residents further validates this socio-spatial bifurcation. Residents from the east stayed mainly within the Mid-Wilshire area for their everyday needs, while many residents in western Mid-Wilshire who owned cars travelled to different parts of metropolitan Los Angeles on a regular basis, such as to Hollywood, Glendale, Eagle Rock, Santa Monica, and even to Monterey Park located fifteen miles (twenty-four kilometres) away to the east. There were, however, a small handful of western residents who regularly made use of the ethnic supermarkets located in eastern Mid-Wilshire as well as public facilities like the LA City College and post office. Overall, the routine geography of easterners and westerners had very little overlap apart from sharing the amenities at Wilshire Branch Library, Burns Park, and the main street of Larchmont Avenue. Figures 5.8 (eastern) and 5.9 (western) demonstrate mobility patterns in opposite directions from Western Avenue.

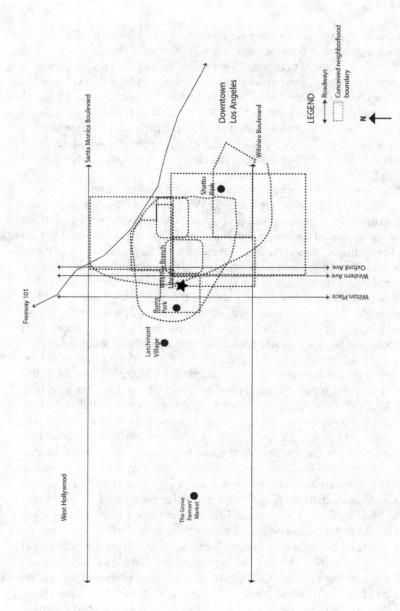

Figure 5.6. Map showing the conceived neighborhood boundaries of eastern Mid-Wilshire residents. Map is not to scale.

Prepared by author.

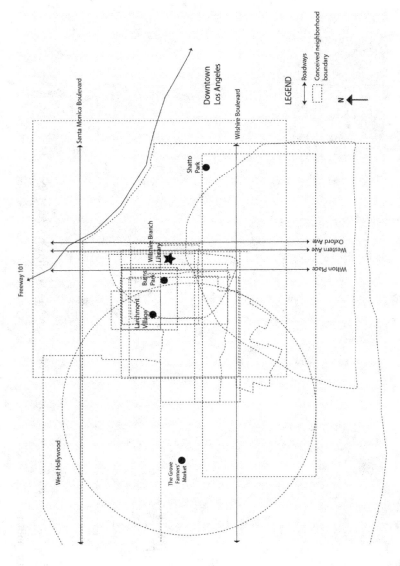

Figure 5.7. Map showing the conceived neighborhood boundaries of western Mid-Wilshire residents. Map is not to scale.
Prepared by author.

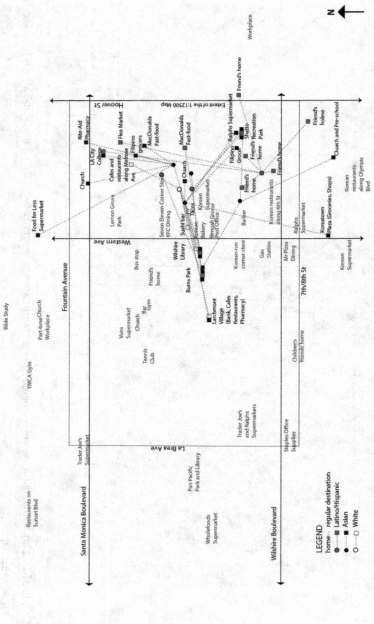

Figure 5.8. Mapping the routine geographies of eastern Mid-Wilshire residents. (N=8). Map is not to scale.

Prepared by author.

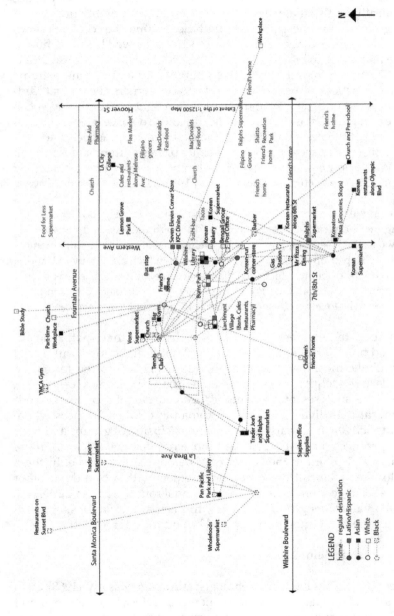

Figure 5.9. Mapping the routine geographies of western Mid-Wilshire residents. (N=14). Map is not to scale.

Prepared by author.

The conceived neighbourhood boundaries reveal how inhabitants of a "super-diverse" locale like Mid-Wilshire imagined social space symbolically and spatially. First, there are multiple neighbourhoods of varying sizes with overlapping boundaries within the locale. Second, beneath a dazzling veneer of diversity, residents viewed Mid-Wilshire as a space marked with boundaries of social, economic, and cultural differences. The diverging set of conceived boundaries and routine patterns of spatial practices between easterners and westerners shows that Mid-Wilshire is not only a collection of multiple neighbourhoods and social enclaves, but more starkly, a symbolically divided space. Depending on which side of the track one lives in, the experiences, practices, and conception of social space differ. Mid-Wilshire brings to mind a description of the West End neighbourhood of Boston before its redevelopment, by sociologist-planner Herbert Gans ([1962] 1982, 11) in *The Urban Villagers*,

> To begin with, the concept of the West End as a single neighborhood was foreign to the West Enders themselves. Although the area had long been known as the West End, the residents themselves divided it up into many subareas, depending in part on the ethnic group which predominated, and in part on the extent to which the tenants in one set of streets had reason or opportunity to use another. For example, the social distance between the upper and the lower end was many times its geographical distance.

Therefore, these three case examples illustrate that spatial neighbourhood boundaries are place-making devices embedded with social and symbolic meanings that can provide insights into the formation of a collective identity and collective life in diverse locales, or the lack thereof. The analyses of residents' conceived neighbourhood boundaries reveal that diverse locales are dynamic territorial spaces where residents actively negotiate inclusion and exclusion of groups and their spatial practices, i.e., who belongs, and what belongs. As such, these boundaries also hint at the contours of belonging in a locale. In these three locales, sociocultural diversity interacts with a bundle of other factors – socio-economic differences, personal routines and experiences, symbolic values, length of residency, municipal regulations, etc. – to influence the conception of the spatial boundaries of the 'hood.

Fostering Local Belongings

The need for belonging, as psychologist Abraham Maslow (1968) theorized, is one of the basic human needs alongside safety, love, relations, and respect that can only be satisfied by an external reality, i.e., other people. We speak of a person belonging to some group or some place. Belonging thus requires "an act of self-identification or identification by others, in a stable, contested or transient way," according to Yuval-Davis (2006, 199). In locales with diverse group affiliations and values, questions about the kinds of belongings that are present and how local belonging is formed are critical in understanding the possibilities for collective life in these landscapes. How does an individual or group form a sense of local belonging in settings like San Marino, where different sets of values and lifestyles are present? Is a sense of local belonging possible in Central Long Beach, where poverty interacts with decades of ethnic gang violence to create exclusions and divisions among people? What kind of local belonging can we expect in settings of multiplicity and inequality, such as Mid-Wilshire?

In the writings about how local belonging forms, there are several common threads as I have outlined (Chan 2013b). First, belonging is a choice. Savage, Bagnall, and Longhurst (2005) conceptualize this choice as "elective belonging," in which the middle-class articulate the right to move and settle in a place that they accord functional and symbolic meanings. The authors found that in a globalizing world, having a sense of belonging to a locality has become more pronounced and important as footloose people choose or "elect" to belong to a neighbourhood and community that they identify with. This notion is shared by Fenster (2005, 227), who writes that "the more choice people have the stronger their sense of belonging becomes." Important to note is also that this choice to belong locally is a form of place belongingness that Antonsich (2010) quoting Yuval-Davis (2006) describes as a feeling of being "at home." Second, belonging is a routine practice and freedom to use and inhabit urban spaces, according to Fenster (2005). Third, belonging for migrants and minorities is usually formed through "their own engagement with the places where they live" and "reflect a perception of being accepted by the majority," as Devadason (2010, 2954) suggests. Fourth, belonging can also be formed by the sharing of ethnic identity and common cultures, as exercised through the definition and maintenance of social boundaries over time (Barth [1969] 1998). Therefore, belonging is a mental conception and an experience. It is symbolic, and it involves a practice that is social and spatial.

In this study, a sample of forty-nine residents who were able to stay to complete the full interview in San Marino, Central Long Beach, and Mid-Wilshire were asked if they felt that they belonged in their neighbourhoods (see table 5.1). About half of the residents who responded to the question in each neighbourhood felt a sense of local belonging. A higher proportion of residents in Central Long Beach felt that they belonged in their locale than those in Mid-Wilshire or San Marino. About 30 per cent of the residents felt a sense of selective belonging – feeling like they belonged in the locale in some respects but not in other aspects. Only a small minority (14 per cent) expressed a definite "no" to this question.

Table 5.1. Reasons for the formation of a sense of belonging or lack thereof in the three locales.

Do you feel like you belong?	Central Long Beach (CLB)	Mid-Wilshire (MW)	San Marino (SM)	Total	Reasons given by participants
Yes	10	9	9	28	• Interpersonal relations with neighbours • Sharing ethnic commonalities • Having local knowledge of neighbourhood • Participating in the community • Feeling a sense of home • Having the right to belong • Access to public spaces and amenities • Sharing same values as neighbours • Living or working in neighbourhood • Proximity to friends and families • Owning property • Birthplace • Choosing to belong
Yes and No	2	6	6	14	• Sharing ethnic commonalities • Presence of ethnic differences • Wealth differences are too stark • Language barriers with neighbours • Different demographic characteristics • No community engagement
No	2	3	2	7	• Poor relations with neighbours • Different values of living • Too many ethnic concentrations
Total	14	18	17	49	–

This table is an adapted version of the original table first published in Chan (2013b).

Cultivating Interpersonal Relations with Neighbours

The analysis of the results shows that for residents in all three locales, having interpersonal relations with neighbours is an important factor that shapes the formation of their sense of belonging in the locale. Conversely, not having interpersonal relations with neighbours is also the major reason for not feeling that they belong. Maintaining good interpersonal relations with neighbours is particularly critical to the formation of belonging for newcomers and immigrants, particularly in San Marino and Central Long Beach.[2]

Among the residents in San Marino who felt like they belonged, they explained that having friends, family, and good neighbours around were important to putting down roots in a place. Two residents whom I spoke to related transformative experiences of how cultivating good interpersonal relations with their neighbours had made them feel included and gave them a sense of community in San Marino. Take for example, Asian American Naomi Su, who had lived in different cities around the world before arriving in San Marino. She shared candidly that growing up in New York City, she never thought that neighbours were people who cared or that she would have a need of them. But when she arrived in San Marino, Naomi was surprised by the warm welcome from her neighbours, who befriended her and helped her family settle in. That experience was so powerful and inclusive for Naomi that she felt for the first time that she belonged in a locale, and more so in San Marino than in any other previous place that she had lived in in the United States.

Similarly, Noelle Lu, a single mother of two and a new immigrant from China who had moved to San Marino not too long ago, experienced warm neighbourly care from the elderly White American couple on her street. She described fondly her neighbours who took the time to engage her in conversations about Chinese culture and her new life in the United States. For Noelle, her neighbours' act of personal kindness and continuous friendliness had been critically important in helping her counter the other bad experiences of discrimination that she and her children faced in San Marino due to cultural and language differences.

The experiences of Naomi and Noelle demonstrate that forming a sense of belonging in a locale requires relational development among neighbours that begins by being accepted and then actively included into the local community. What is even more exceptional in their case is that good interpersonal relations with neighbours had not only grown in them a sense of belonging in San Marino but had inspired them to deepen their identification with the locale. They became active volunteers in the public schools and in other civic programs so that they could help to make San Marino a place that more residents could call home. The respect, warmth, and acceptance that residents receive from their neighbours is critical in nurturing a sense of local belonging, particularly among the foreign-born and minority population, which in turn builds new social capital for a neighbourhood.

Ahn, a Vietnamese immigrant who had owned and operated a business in Central Long Beach for twenty years, said that working in the neighbourhood for so long meant that she had been able to get to know and earn the trust of her multi-ethnic customers and their families. Her customers had developed familiarity with her so that she had become a confidant to many of them. These relationships had developed her strong sense of belonging and attachment to the locale. Aside from Ahn, the small minority of White residents whom I interviewed also expressed the same sentiments that good interpersonal relations with neighbours was extremely important for them, as far as feeling like they belonged in a locale. For example, Rich Taylor, a chatty White American man in his late fifties whom I met in the Mark Twain Library, relocated to Central Long Beach two years ago. He described his friendships with his Cambodian neighbours as the major reason for his sense of local belonging. He said that his Cambodian friends had helped immensely in getting him to be a part of the social life in the area quickly. He spoke with pride about how he regularly hung out with his Cambodian friends at the local Cambodian cafe:

> I am accepted by the Cambodian community – all facets of it. I am like a son almost to a lot of them ... The Hispanics know that I am a cool guy and Black people know I am a cool guy. All kinds of people who know me cover for me. It is pretty okay for me to go anywhere.

From the information gathered from the interviews, good interpersonal relations with neighbours are not only symbolic and emotional connections but are a means to access and have good local knowledge of the neighbourhood. This knowledge includes basic wayfinding, information about community events, knowing who does what, and having a good sense of the resources and agencies in the locale. It is no surprise that community organizers and shopkeepers in Central Long Beach felt that their sense of local belonging is formed through their engagement with the diverse groups of local residents and gaining their trust as familiar representatives among them. In San Marino, local knowledge of the neighbourhood is also learned from participating in neighbourhood activities such as fundraising, clean-up, and school programs. A resident gains social exposure, increases contact with neighbours, and earns social acceptance. In a city where sacrifices made for children's education and volunteerism are valued more highly than further accumulation of wealth, gaining knowledge through participation in the community is especially significant to foster a sense of belonging. In fact, in San Marino, there is a tacit expectation for its residents to engage in civic matters as a measure of belonging to the city.

Sharing Ethnic Commonalities

In Central Long Beach, residents felt that the presence of co-ethnics was very important to the formation of local belonging, particularly the young second-generation Latino adults. Similarly, long-time Cambodian immigrants explained to me that their sense of local belonging was built on the familiarity with the social environment of Central Long Beach because it was one of the biggest (if not the biggest) clusters of Cambodian commerce and culture in the United States. This locale is a gathering place to socialize with co-ethnics, some of whom travel long distances from elsewhere in Southern California to meet here. The Cambodian restaurants in the area are venues for major community events, weddings, birthdays, and more for the Cambodian diaspora. The presence of co-ethnics of equal socio-economic status offers many more possibilities of forming bonds of belonging in a locale as shown by Lee's (2019) research on Los Angeles's middle-income Asian and Latino homeowners. For example, these locales are home to residents that are more likely to share common social and cultural values rather than conflicting ones, and offer the benefit of amenities and third places of social gathering like ethnic grocery stores and restaurants, as shown in the case of Central Long Beach for the Cambodians. However, while local belonging formed by ethnic ties can be a boon for social bonding, it can also create social, spatial, and symbolic boundaries among groups that do not share a common ethnicity. These boundaries have the potential of fragmenting and dividing a demographically diverse locale socially.

Similarly, the residents' responses in eastern Mid-Wilshire indicated that sharing ethnic affiliation was a double-edged sword with respect to formation of local belonging. Korean, Hispanic, and Filipino residents told me that their sense of local belonging was an outcome of the proximity and presence of co-ethnics, including those of the second generation. In Luciana Garcia's symbolically charged description of walking home in a locale of diverse ethnic groups quoted earlier in the chapter, we are reminded that belonging is circumstantial and dependent on an external reality as Maslow (1968) observes. Yumi Lee, a first-generation Korean immigrant who married a Korean American man and moved to Mid-Wilshire about five years ago, felt a mixed sense of belonging in Mid-Wilshire – one that was significantly defined by her ethnic affiliation as a Korean living in Koreatown. "Do you feel like you belong in this neighbourhood?" I asked.

Half and half. I live here. Because my husband's friends are born here but I sometimes misunderstand them [because of English]. I belong because many people speak Korean in LA.

In contrast, in western Mid-Wilshire, it is the lack of co-ethnic presence (or more correctly, the presence of ethnic differences) that have helped a small handful of residents foster their sense of belonging. Larry Gans, a White American homeowner, loved the "wonderfully weird" and the "eclectic" diversity of Mid-Wilshire. There was also Nancy Lau, a Chinese American who remained fascinated by the diversity in Mid-Wilshire even after thirty years. For them, they had elected to belong to Mid-Wilshire, sharing a similar kind of belonging as many residents in San Marino. In Nancy's words,

I think LA is the best example of what diversity can look like. Not just physical appearances but to be more open about embracing the culture. You can't avoid it. You drive by and you see the different kinds of fruit from South America. How by adding a little brown sugar, it changes the flavour … It is just beautiful … Being different is not bad … You got to be open or else you will become isolated. Soon it becomes more difficult for you to live, especially when you identify the Mid-Wilshire area. You see all of that. You go to Alvarado and Sixth, you see the pushcarts and variety of food in "pop-up" mom and pop restaurants. You go to LACC [Los Angeles City College], you see Armenians, East Indians. It is great … we just love LA. We really do.

(S)elective Belonging

Gathering the different perspectives of how local belonging is formed in the three locales, I observed that in the presence of demographic diversity, belonging is always partial. Even in circumstances where residents have the resources to choose to belong and possess a sense of elective belonging, their sense of belonging is selective at best.

For example, in San Marino residents have bought into the locale for its values (both pecuniary, in terms of property values, and philosophical, in terms of the focus on education and family that the city offers), and they demonstrate a sense of elective belonging per Savage, Bagnall, and Longhurst (2005). However, exercising the choice to belong and expressing it through engagement in local activities may not be sufficient to give one a full sense of local belonging in San Marino. In the interviews, six residents communicated a sense of ambivalence in their local belonging there. Five of them were first-generation immigrants from Taiwan and China. These residents had *elected* to belong in San Marino because they shared its value of being a safe space to raise families, and as part of this election, they had chosen to become active volunteers in local fundraising events. However, they confided that on a deeper level, they did not feel like they fully belonged in and were fully accepted by the White American community in San Marino because of certain unpleasant experiences arising from language differences and cultural preferences.[3] This was a stark contrast to the White American residents, who had quick and affirmative responses when asked if they belonged in San Marino. Therefore, formation of local belonging cannot be purely a choice, as if no barrier exists, because it always requires negotiation of barriers and social boundaries that create inclusions and exclusions.

Where elective belonging is observed in these locales, a pattern of *selective* belonging is also present. It is common that among the Chinese immigrants in San Marino, including those of the second generation, many do not feel that they belong entirely in the neighbourhood because of their immigrant status, ethnicity, linguistic ability, and family background. Thus, they share a sense of ambivalence and alienation that manifests in these feelings of selective exclusion from the community. Bulgarian-French philosopher Julia Kristeva (1991, 15) described this sense of ambivalence well in her nuanced interpretation of the "floating foreigner":

> No one points out your mistakes so as not to hurt your feelings, and then there are so many, and after all they don't give a damn. One nevertheless lets you know that it is irritating just the same. Occasionally, raising the eyebrows or saying "I beg your pardon?" in quick succession lead you to understand that you will "never be part of it," that it "is not worth it," that there, at least, one is "not taken in." ... Thus, between two languages, your realm is silence.

In western Mid-Wilshire, we see another form of elective and selective belonging that is built upon a defensive NIMBYism mindset, likely due to its locational proximity to low-income neighbourhoods. A sense of "elective" local belonging is evident through neighbourhood activism that organizes and protects the interests and amenity of residents. These neighbourhood coalitions, according to Charlie Brooks, an African American man in his thirties who had moved into the area a few years ago, were extremely vocal and powerful when "their boundaries are threatened." Charlie's choice of the word "boundaries" reflects an important and sensitive issue in this densely populated and polarized part of Los Angeles – the safeguarding of privacy, safety, and neighbourhood amenities against trespassers. At the Greater Wilshire Neighbourhood Council meeting that I attended in 2011, residents rallied around issues of neighbourhood aesthetics, traffic, and safety. These motivated residents scrutinized every new development proposal, commenting on their form, architecture, traffic impact, and safety implications. Nancy, a resident in the wealthy Hancock Park area, succinctly summed up the territoriality and elective belonging in this locale:

> In this particular area, there is still the influence of the well-to-do. They still have a lot of power to control what goes on … When one calls the councilman, the councilman really responds. The neighbourhood still holds its clout in terms of the resources it can command.

Elective belonging practiced here shares traits with British geographer Paul Watt's (2009) "spatially selective version" of belonging. In his study of the middle-class English suburb of Eastside in London, he found that new middle-class home-owning residents *selectively chose* to associate their pocket neighbourhood of Woodlands with a place image of suburban exclusivity and to disaffiliate from the rest of Eastside, which is an area with large tracts of council housing. According to residents like Mark Adams, a Black American hair stylist, and Matthew Cruz, a Filipino American studying to become nurse practitioner, what distinguished western Mid-Wilshire from its surroundings was the professionalism demonstrated by its residents. "They know how to act and carry themselves," said Mark. "They look all stable, are educated, self-reliant, possess a work ethic and have a FICO[4] score of 702!" Mark believed that professionalism enlightened one's mind to respect the individual and their privacy, and inculcated tolerance of different cultural habits. Mark gave this example: even though his Jewish, Asian, and Indian neighbours were inward-looking and did not seek to befriend or engage others, a person of professionalism like himself would not take offence and would remain a good neighbour.

Another form of selective belonging is observed among a few (about 12 to 16 per cent) residents who elect *to not belong* because they feel that the values of the neighbourhood do not reflect theirs. Often, these residents experienced difficult interpersonal relations with their neighbours, or a lack thereof. In eastern Mid-Wilshire, first-generation Filipino immigrant Chloe Castillo said that she wanted to belong in a different, calmer area, where neighbours were more respectful. Her current neighbours were loud and rude, displaying a lack of care for the community. Chloe felt that the neighbourhood was no place to raise a child. In Central Long Beach, Black Americans like Calvin Jenkins (a new resident) and Marteese Owens (a long-time resident), felt alienated and unhappy because of the undercurrents of antagonism between Blacks and Latinos in the locale. In fact, Calvin told me, "I am serving time here!" Similarly, Marteese, who had lived in Central Long Beach for most of his life, spoke about the difficulty of developing a sense of belonging in a place where his values of social life no longer fit with those around him.

MYSELF: Have you ever felt like an outsider in this neighbourhood?
MARTEESE: All the time. You are taught to stay away from certain people because you don't want a conflict. Instead, you tend to stay around your own, in your own little area, in your own cubicle.
MYSELF: Who are these people? Is it ethnicity or race?
MARTEESE: No. It is the ignorant mentality of certain people. It is not really a race. They always have to cause problem in order to feel good.
MYSELF: Do you feel like you belong?
MARTEESE: I don't think so. I think I should be somewhere else.

Marteese went on to tell me that once he was done with college next year, he would be relocating with his wife and children to Alabama. In fact, his elderly parents had made the move earlier and liked it there. He had outgrown the locale and longed for a place that was different, where people were less ignorant.

Concluding Thoughts: Crossing Boundaries

Through the discussion in this chapter, I have illustrated that the social space of diverse locales is embedded with visible and invisible boundaries that have spatial, symbolic, and social characteristics. Some boundaries are collectively recognized and physically visible, while others are individualized and less palpable. In neighbourhoods of homogeneous income levels such as San Marino and Central Long Beach, residents share a collective set of conceived spatial boundaries of their neighbourhoods. In contrast, Mid-Wilshire's income levels are mixed, and residents' conceived neighbourhood boundaries vary in size and reach. In western Mid-Wilshire, wealthier residents' mental images of their neighbourhoods have extensive geographical limits, while residents in lower-income parts of Mid-Wilshire visualized their neighbourhoods as small areas with well-defined boundaries. This could be due to the freedom of mobility enjoyed by higher income residents that comes with automobile ownership, which expands their geographical range beyond the limits of the immediate surroundings. While there is no common set of neighbourhood boundaries in Mid-Wilshire, there is, however, a collective conception by residents from eastern and western Mid-Wilshire of a strong spatial, social, and symbolic boundary that runs through the locale along Western Avenue and cuts Mid-Wilshire in half.

These boundaries of diverse locales are dynamic and meaningful because they reveal the geography of inclusion and exclusion that shapes the formation of local belonging between social and cultural groups. For example, in San Marino, the clear match between the residents' conceived spatial neighbourhood boundaries and the official municipal boundary vis-à-vis visitors to the area reflects a strong sense of collective, elective belonging to a wealthy and well-run locale among the residents. In the case of Central Long Beach, many residents *and* visitors to the area share a common set of neighbourhood boundaries that indicates a collective subconsciousness of the neighbourhood as having a different character than its surroundings – a population-dense and worn-looking neighbourhood that is juxtaposed with the spaciousness and glitter of the rest of City of Long Beach. This collective conceived image interacts with the daily spatial practice of circumscribed mobility in the locale by its residents and visitors alike, reinforcing an enclave identity. These physical and social boundaries are further bolstered by a strong presence of co-ethnics in the neighbourhood to produce a sense of local belonging felt by its residents. In Mid-Wilshire, the strong spatial boundary along Western Avenue conceived and perceived by residents from both sides points to the felt and visible presence of income segregation that has made belonging highly selective among its residents.

Crossing spatial, social, and symbolic boundaries in these landscapes of social and cultural differences is very much part of an everyday spatial practice for most who live in, work in, and regularly visit these locales. For residents who live in eastern Mid-Wilshire and Central Long Beach, boundaries are made palpable by signages of different foreign scripts on storefronts that are interspersed throughout the urban landscape, and gang territories are made visible by gang-tagging graffiti markings and street violence. Be it driving through a street block occupied predominantly by one social group, or walking along a street lined with different foreign businesses and languages, the experience of boundary crossing subjects one to feeling like an outsider in one's 'hood. This feeling is frequent, routine, and visceral. Yet, despite these daily crossing, actual cultivation of intercultural relationships is rare. This is not to say that without these palpable symbols of boundaries, as in the case of San Marino, which has stringent regulations against large foreign script signages, intercultural relationships would be easier to form.

Boundary crossing takes effort, time, and skills. It requires knowledge of the types of boundaries that are present and discernment of the right time to make the crossing. It has its risks and rewards. Living in a diverse locale can be difficult for those who are not ready to negotiate the boundaries. Residents in eastern Mid-Wilshire and Central Long Beach repeatedly alluded to fear and insecurity in navigating differences that come with boundary crossing in spaces with multiple overlapping territories that are fluid and dynamic. For example, a public park could become a secondary or even primary territory of certain groups, depending on how explicitly or tacitly these groups limit the access to the public space for others.[5]

Amorphous territoriality intensifies the anxiety of boundary crossing, in addition to that caused by the lack of capability to cross language boundaries. As Stanley Milgram (1970, 1463) explained, in addition to the stimulus overload coming from the bombardment of linguistic, cultural, and social differences, and expectations that characterize life in cities, "diversity also encourages people to withhold aid for fear of antagonizing the participants or crossing an inappropriate and difficult-to-define line."

In comparison, boundary crossing is a more managed experience for the residents in San Marino and western Mid-Wilshire. This is because boundaries in these locales do not always have to be crossed. Their residents have the resources and capabilities to circumvent these boundaries or exercise consent to non-encounters. For example, travelling in a private automobile, as compared to taking public transportation and walking, an individual can go directly to the destination without any interpersonal encounters with others. Thus, boundary negotiation is selectively entered into by those who have the social, cultural, and economic capital to do so. Where boundary crossings are made, they are usually strategic as well as tactical to the interests of the parties involved, such as volunteering in children's school activities, joining social club events, and participating in neighbourhood advocacy groups.

Boundaries are double-edged swords that have implications for the formation of local belonging. On the one hand, boundedness creates safety, stability, and belonging that can empower individuals to venture a daily or future crossing – akin to how a stable home environment can empower an individual with the confidence to step out and take on new adventures. On the other hand, they are powerful tools of social, physical, and symbolic exclusion. Likewise, boundary crossing promises both productive and unproductive outcomes. According to Geraldine Pratt (1998, 35), "Boundary crossings can also disempower, fragment identity, and protect privilege, and bounded communities may have progressive effects." The process of crossing destabilizes the status quo. Boundary crossing could agitate and exacerbate dualistic tensions if the wrong ones are crossed in the wrong way, at the wrong time or place. However, if done right, boundary crossing could unlock dialogical tensions that have the potential of bridging differences.

In diverse locales where residents have an interest in cultivating collective life, an awareness of how social boundaries interact with and articulate in space is critical because of the salience of territoriality in these places. The management of boundaries in diversity is challenging and requires determining which set of boundaries is elastic enough to become more inclusive, and which set to keep at status quo because it is core to the social stability of the locale.

6 Intercultural Contours of a Diverse Public Realm

It is all cultural, and you just adapt to the culture of the neighbourhood ...
You have to be diplomatic and have to exercise some skills when you are not
familiar with culture. That is all these communities are about – culture. It is just
about learning from each other and figuring how we are going to get along,
and what are the concessions that we can make to live in peace and harmony.

Tania Johnson, Black American
in her thirties who lives in Mid-Wilshire

Demographic diversification destabilizes and complicates the public realm
of diverse locales as constant consternation about the unexpected and the
strange undermines the formation of interpersonal relations that are critical
in the fostering of local belonging. Sociologist Lyn Lofland (1973) described
the public realm of big, dense, and diverse twentieth-century cities as effectively a "world of strangers" where anonymity exists alongside familiarity. Multiple and overlapping social boundaries fill the relational web of
the public realm with complete strangers and "familiar strangers" whom
we recognize categorically but never get to know personally, according
to social psychologist Stanley Milgram (1972). The twenty-first-century
metropolises are bigger, likely denser, and certainly more diverse due to
widespread global immigration. The city is an ever-expanding world of
strangers that make intercultural living ever more challenging.

In this chapter, I will gather the strands of discussion about the spatial
practices, tensions, and belongings in diversity to reconstruct the relational web and trace out the intercultural contours in the public realm of
diverse locales by asking these questions: How do inhabitants in these
diverse locales view and understand interculturalism as an everyday
lived experience? What are the barriers to intercultural learning and
understanding, i.e., the conditions that keep the public realm of diverse

locales as a space of familiar strangers? And what are the opportunities for interculturalism in the local public places of gathering?

Configurations of the Relational Web

The interviews with residents revealed that the *relational webs* in the public realm of these locales of diversity in Los Angeles are comprised largely of *fleeting* encounters between complete strangers and *routinized* contact experiences, such as the brief hi-bye exchanges between categorically known neighbours (Lofland 1998).[1] In the case of Central Long Beach, where residents, vulnerable to rent hikes, change apartments frequently, the transience of their living location further deters investment in relationships. In addition, poverty, the lack of a common spoken language within the neighbourhood, and the constant fear of becoming a victim of street violence has resulted in the residents keeping to themselves and minimizing public engagement.

The public realm in these diverse locales can, under certain conditions, offer *quasi-primary* contact, i.e., encounters that last from a few minutes to hours between people who categorically know each other. Neighbours who have repeatedly seen each other on the sidewalks, at the playgrounds or libraries, in neighbourhood shops, or have other social spheres that intersect would engage in longer conversations about their children, parenting and work challenges, and common issues that affect the neighbourhood. Thus, some level of social trust could be established in diverse locales from familiarity alone, although not adequate.

Between complete strangers, LA small talk is a commonly used engagement mechanism to establish a *quasi-primary contact* experience by eliciting and exchanging personal information about each other, about their places of origin and years of residence, and at times, even fashion tastes. In densely populated and economically disparate Mid-Wilshire, where a high degree of urban anonymity has created a counter desire in some residents to know others and be acknowledged by them, LA small talk as a mechanism to establish contact with strangers is frequently employed. As sociologist Erving Goffman (1963) underscored when concluding his study of *fleeting* and *routinized* contact behaviours between people in public, passivity should not be dismissed as non-interaction. Although LA small talk is often belittled for its inability to make durable relations, its frequent use as a border-crossing technique by individuals in diverse locales to craft dialogical tensions in the face of dualistic forces conveys the salience of daily tactics to maintain coexistence in socially complex environments.

Experiences of long-lasting *intimate-secondary* emotional relationships in the public realm between individuals from different ethnic groups were uncommon in all three locales. These *intimate-secondary relations* were usually found in social clubs where common interests and goals transcended differences, such as parent-teacher associations, and the Rotary Club of San Marino, or between co-organizers of neighbourhood activities in Mid-Wilshire. However, according to the participants in San Marino, these *intimate-secondary relations* were difficult to sustain over time as shared purposes and interests dissolve, for example, due to children graduating from schools or reaching a conclusion in event planning.

The relational web in the public realm of diverse locales must also be read in the context of the other realms of city life, such as the parochial realm of the neighbourhood and private realm of the home (Lofland 1998).[2] In the three locales, the presence of *comfort zones* was prevalent. *Comfort zones* include places outside the home: for example, shops or restaurants that an individual frequents because he or she feels comfortable there and can communicate with the service staff or access products that are familiar. *Comfort zones* can also be set paths an individual takes in his or her routines. For example, one participant, a Cambodian who had lived in Central Long Beach for more than ten years, would only take the same exercise circuit route every morning because she felt safe and familiar with the environment. In cases like this, *comfort zones* behave as parochial spaces that exist within the larger public realm to counter the discomfort of the unpredictability of the public realm.

Comfort zones are mental, spatial, and symbolic spaces of familiarity, stability, intimacy, safety, acceptance, and refuge, which have the potential to become social spaces that are highly selective and exclusionary; thus, they have more characteristics of the private realm rather than that of the public realm. Once formed, these cocoons of comfort are hard to penetrate, even for newcomers who share the same language and cultural background. In this way, comfort zones curtail the development of the public realm in diverse locales.

In this study's three locales, several kinds of *comfort zones* existed, in particular those that drew on ethnic or nationality ties. For example, in San Marino, many of the first-generation Chinese residents indicated their *comfort zones* to be occasions, group settings, or places in which they were able to use Mandarin to socialize with other individuals who had similar life experiences and backgrounds, such as fellow immigrants from the same city or university. Sandy Cheng, a first-generation self-identified Asian/Chinese American who had lived in the United States for about forty years and spoke eloquent English, shared her take about whether Chinese residents had adapted to the American culture in San Marino:

> I really don't think, probably a handful of people are really assimilated. People all have their *comfort zones*. Even as assimilated as myself, I prefer

to speak Chinese, to eat Chinese food, and to watch Chinese TV. I can do both. I enjoy both. But when it comes to choices, I would choose a Chinese restaurant 90 per cent of time over Western food. But the second generation is probably quite assimilated. It really depends on the family. Some who move there would never speak a word of English, who watch Chinese TV program, who read Chinese newspapers, go to the Chinese grocery to shop, and they speak Chinese only in the house.

Similarly, in Mid-Wilshire, second-generation Hispanics and first-generation Korean immigrants described their comfort zones as spaces of homogeneous social and cultural practices, i.e., ethnic bubbles as comfort zones. In contrast, comfort zones in western Mid-Wilshire were spaces of cultural heterogeneity but also socio-economic homogeneity. Many of its residents had elected to live in and belong to this culturally diverse part of Los Angeles because it offered a diverse environment without the discomfort of income disparity.

One's *comfort zone* can be another's discomfort zone. For many of the White American residents who have lived in San Marino for over thirty years, the entire city used to be their *comfort zone*, when it was less ethnically diverse. As San Marino diversified socioculturally, the *comfort zones* of some residents shrunk as the social space became less familiar and social tensions arising from differences disrupted the feelings of stability and social acceptance. The prevalence of these ethnic *comfort zones* in these diverse locales has the tendency to undermine the development of a viable parochial realm of the neighbourhood based on proximity between residents, and the development of a broader public realm that enables intermingling among social and cultural groups.

Interculturalism in Los Angeles

Much of interculturalism research has focused on formal dialogues among groups rather than the conception and practice of intercultural relations by people who live in diverse locales. In this study, sixty-eight participants in the three locales responded to the following question, "If you have the opportunity to meet and get to know a neighbour of another ethnicity or nationality, what would you like to know about him/her?" Most of the participants, who came from different socio-economic backgrounds, felt that learning who their neighbours were, what they did, and their reasons for living in the neighbourhood were most important – interpersonal knowledge was a clear priority. Learning about and understanding their culture and customs was the next most important. See figure 6.1 for the comparison of responses in the three locales.

Figure 6.1. What would you like to know about a neighbour of another ethnicity or nationality?

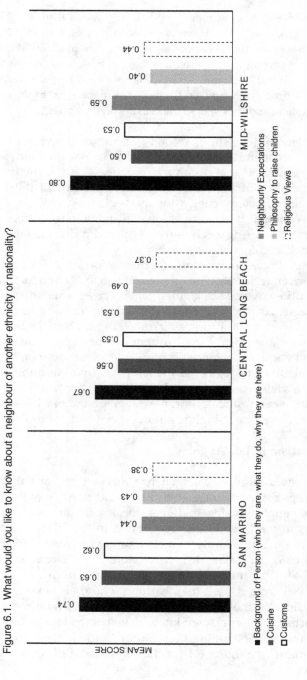

(N=68). Mean Score of 1.0 indicates most important.[3]

Table 6.1. Four major definitions of intercultural understanding given by a sample of participants

1	Gaining a deeper and more empathetic understanding through knowledge of a person's background, personal values, and what drives them, such as their dreams and disappointments (e.g., how their kids are being raised, why they do what they do, their perspectives on social issues).
2	Learning to appreciate different aspects of cultures and their practices (e.g., cooking styles, cuisines, traditions, customs, language, habits, music, drama, and literature) as well as accepting and adapting to them.
3	Recognizing our common humanity, needs, and values, looking out for each other, and working together.
4	Having tolerance.

The importance of interpersonal knowledge in diverse locales emerged again when residents, municipal officers, and community organizers were asked for their conceptions of intercultural understanding. For many of them, intercultural understanding was a process of gaining a deeper and more empathetic understanding of another person from a different social and cultural group by learning about the cultural context that shaped their thoughts and beliefs, such as customs, language, and values. See table 6.1 for the four major definitions given by the participants in descending order of frequency mentioned.

Gaining a deeper and more empathetic understanding through knowledge of a person's background provides a foundation for the cultivation of future friendships. The participants who viewed intercultural understanding this way were mostly first-generation immigrants and second-generation Mexican Americans. Their orientation to this definition of intercultural understanding could very well be informed by their mixed and hybrid identities that resisted easy categorization along the lines of ethnicity or nationality.[4] Because of their background, they could have developed a greater sensitivity in approaching cultural differences. A second-generation Mexican American resident in eastern Mid-Wilshire said,

> For me, intercultural understanding means what we could possibly do to bring different ethnicities and races together by understanding their cultures and their differences, and to at least have a common understanding between each other so that we won't have stereotypes come in between us.

There was also a sizeable group of participants for whom intercultural understanding was learning to appreciate and gain knowledge of different cultures, including cuisine, customs, and language. A few people in this group believed that learning should not remain purely cognitive but required acceptance and adaptation. For them, intercultural learning and understanding meant integrating and blending in because living peacefully in diversity requires that. Nick Chang, a first-generation Taiwanese American, described the process of intercultural learning:

> You educate them about your culture, and you learn their culture. We can do that by exchanging experiences of our education and our daily activities. As the Chinese saying goes, 入乡随俗 [ru xiang sui su]. In other words, when you are in a new place, you learn their culture, accept it, and blend into it.

A much smaller group of participants thought that intercultural understanding was a bridge to recognize our common humanity across cultural differences. Jonathan Anderson, a White American in his twenties from Central Long Beach, defined intercultural understanding: "It is people getting along, people understanding each other, people of different races uniting, looking out for one another." For this group of participants, how we look and behave are merely veneers on a core of universal human needs, motivations, and desires. As Tania Johnson, a Black American resident in Mid-Wilshire, explained, "We have a lot in common and share the same values such as opportunities for our children, healthcare, education, etc." Damien Torez, a community organizer in Mid-Wilshire, had come to embrace the virtue of multi-ethnic community participation as a means by which different social and cultural groups could develop camaraderie, and had come to this realization: "We are in the same boat." For this group of participants, intercultural learning and understanding is about recognizing and embracing similarities over differences.

A couple of participants felt that achieving intercultural understanding is too high a bar; achieving tolerance is a more suitable outcome to expect. They explained that tolerance is a form of respect for different and strange practices, which could open doors for subsequent engagement and understanding; thus, tolerance rather than understanding should be the goal in diversity. On this point about tolerance, some participants felt that the concept has a negative notion and opposed to tolerance as a prerequisite for intercultural understanding. Kelly Douglass, a community organizer from Mid-Wilshire, shared a similar perspective with scholars – that tolerance does not actually require respect; rather, tolerance is equivalent to *putting up with* practices that are perceived to be negative until these disagreeable differences dissolve.[5] Intercultural understanding, according to Kelly, should "leave space for expression of difference. It does not require or assume that we are alike in anyway. It is beyond tolerance; it should be enjoyable."[6]

In the formulations of interculturalism according to the inhabitants of these diverse locales, there was a distinctive aspect of practice that informed its meaning for many of them. Intercultural learning is about gaining practical knowledge to live better lives with others who are different in the contexts that come with unequal power relations (e.g., those who speak English and those who do not speak it well) and income inequalities (e.g., in Mid-Wilshire). The emphasis for most of the participants was, in fact, to go past group boundaries to establish interpersonal understanding through exchange, interaction, and mutual learning.

Barriers to Intercultural Learning and Understanding in Los Angeles

An analysis of the reasons given for the lack of intercultural learning and understanding experienced reveals that there were common bundles of barriers faced by individuals across the three locales. These include the lack of openness among residents of different ethnicities and nationalities due to the availability of cultural comfort zones to hunker down in and the presence of language barriers. In addition, the widespread dependence on private automobiles in Los Angeles to conduct daily life for most created a self-isolating lifestyle, particularly for residents in Mid-Wilshire. See table 6.2 for barriers listed according to frequency cited by participants.

Lack of Openness Due to Comfort Zone

A common barrier across the three locales is the presence of the double-edged sword of *comfort zones*. As discussed in the previous section, *comfort zones* established by language and culture can offer support for co-ethnics but also reduce the willingness to learn about another or to share one's culture with another. *Comfort zones* draw similarities in and push dissimilarities out. This observation is consistent with Putnam's (2007, 137–51) findings in ethnically diverse American neighbourhoods that "residents of all races tend to 'hunker down'" and "at least in the short run, seems to bring out the turtle in all of us." That reduction of the need for interaction across differences can over time reinforce stereotypes of outsiders, particularly in locales of already polarizing diversities like Mid-Wilshire, and where years of inter-ethnic mistrust

Table 6.2. Barriers of intercultural understanding in each locale ranked according to frequency cited by participants

San Marino	Central Long Beach	Mid-Wilshire
1. Lack of openness due to comfort zone	1. Lack of openness due to comfort zone	1. LA Automobile Bubble Lifestyle
2. Lack of time	2. Language barrier	2. Language barrier
3. Language barrier	3. Poverty	3. Lack of openness due to comfort zone
4. Lack of community space	4. Gangs	4. Ethnic territories
5. Self-sufficiency	5. Ethnic territories	5. Lack of community space
6. Lack of common interests	6. Lack of trust	6. Lack of time
	7. LA Automobile Bubble Lifestyle	7. Self-sufficiency
	8. Lack of common interests	

bundled with poverty and gang violence have solidified views of "the other," as in Central Long Beach. Marteese Owens, a resident in Central Long Beach, also gave me another perspective on the nuances of how institutionalized racism can reinforce these barriers found in diverse locales among Black, Latino, and Asian residents:

We just need more White people to come to the park and library to hang out because that is keeping everything off balance. Most of the White people who come around are the police. The police sometimes throw their ignorance out there at certain people. They don't like certain people, and so they will mess with the bunch and try to nick pick and see who it is. But they are not around enough to do that to figure out who is really the bad apple. They just pry us. The Asians are not tripping anyone, the Mexicans are not tripping anyone. But now the police are tripping. Why did the police trip? What did y'all do? Now we start looking at each other like what are they messing with us now? If we have more White people probably in there, the police probably would not do that. Because they would be like they are friends, they are not trying to make deals, whatever. They nick pick on us to figure out why we are hanging out. There is always somebody trying to break up the group, so we tend to stay away from the groups. Why try? When we are finally having a good time, the police is trying to mess it up. Blacks and Asians don't fight, even the Samoans. For some reason, I don't know why, it always affects the Mexicans [*lowering his voice*]. The police always influence the Mexicans to turn their backs on us.

LA Lifestyle: Car Culture in the City of Survival

Another notable barrier is the car culture in Los Angeles that has created a lifestyle that keeps people to themselves. The participants spoke about how life in Los Angeles was always on the go and the reliance on private metallic car silos to get around reduced the opportunities for social encounters and interaction among neighbours who lived in the same locale. This phenomenon of car culture and sprawl affecting community life in America is also underscored by Putnam (2007, 213): "the car and the commute ... are demonstrably bad for community life" as "we are spending more and more time alone in the car."

Tania Johnson, an African American female resident in her thirties who relocated from the eastern half to the western half of Mid-Wilshire, told me that although living in the less well-off and less automobile-dependent eastern Mid-Wilshire was culturally isolating because she was surrounded by Spanish-speakers whom she could not communicate with, she at least saw families and people out on the streets talking to each other there. However, in the wealthier western neighbourhood where most residents own personal automobiles, she saw fewer street activities and social interactions between neighbours. People were mostly into individual activities of dog-walking, jogging, or strolling with their children. In fact, as the mapping of the routines of western Mid-Wilshire residents in figure 5.9 illustrates, residents in western Mid-Wilshire regularly travel out of their neighbourhood for their routines.

In addition, Los Angeles is also recognized as a survival city by many of the participants: a place where people have come to seek fame, fortune, and to make a living. Many are working two jobs to make ends meet so there is little time to interact or get to know neighbours and the different cultures they live with. According to Larry Gans, a long-time Mid-Wilshire resident, intercultural learning and understanding from his point of view is an "irrelevant" notion:

> The reality is in Los Angeles, people come here to make money. Pure and simple. The relationships that get established for me are primarily through my children. I have met people through college who have become my friends, but we are like ships passing each other through the night because we are busy doing our own things ... As far as different cultures and stuff like that, I honestly feel that it is kind of irrelevant. Everybody is like moths attracted to the light. Like in Los Angeles, people like the weather and there are a lot of things to do. You know, you are attracted to a place like that where you can have a good quality of life. But as far as whether you are working with somebody who is Hispanic or Asian or Caucasian, who cares? Everyone is trying to survive. That's the reality in Los Angeles to me.

Lacking Community Space? Intercultural Opportunities in the Public Realm

The public realm is formed by different kinds of places of public gatherings. Some *public places* are free to access and can be used by all, such as the public parks, libraries, and recreation centres. Some are *semi-public places* that are open to anyone who will pay a fee to enter and use the amenities, such as cafes, restaurants, supermarkets, corner shops, pubs, bookstores, hair salons, and even schools. Some are *private spaces or club spaces* that require pecuniary or non-pecuniary membership to access, such as country clubs, alumni clubs, and homes of friends. Among these places, some are more parochial than others, i.e., communal and frequented by residents in the neighbourhood. Geographer Ray Oldenburg (1989) termed the spaces where neighbours meet and mingle in a locale as "third places." In contrast to the first and second places of home and workplace respectively, "third places" is "a generic designation for a great variety of public places that host regular, voluntary, informal, and happily anticipated gatherings of individuals beyond the realms of home and work" (Oldenburg 1989, 16).

Although the camaraderie identified by Oldenburg and enjoyed in the third places could have taken place in study locations that are less socially heterogeneous, his finding that additional and alternative places are critical in developing social relations between neighbours is equally, if not more, pertinent for diverse locales. In these locales, there are usually fewer common first and second places, and as the findings presented thus far have shown, there is a need for known and safe places that familiar strangers can mingle and seek out conviviality – places where dialogical tensions can emerge to counter the dualistic tensions that tend to accompany differences.

A mapping survey of places available for social gathering (public, semi-public, private-club, schools, and places of worship) in the three locales reveal that there are indeed very few public places, i.e., free and open to all in terms of access, found within the study boundaries, as highlighted by the residents as a barrier to intercultural learning and understanding. In fact, places of social gathering are mostly semi-public in character, i.e., paid use with open access to all, such as cafes and supermarkets, or private-club spaces, in which access is granted by pecuniary or non-pecuniary membership only. The maps also show the potentiality of school grounds as gathering spaces because of their strategic locations in the residential fabric of the locales and their adjacencies to other public amenities, such as in the cases of San Marino and Central Long Beach. In San Marino, school grounds are already being actively integrated into the daily recreation and

Table 6.3. Different types of gathering spaces in the locales

Access	Uses included
Public places (free and open to all)	Park, library, post office
Semi-public places (paid use and open to all)	Bookshop, cafe, restaurant, supermarket, shopping mall, store/Shop, pharmacy, salon, learning centre
Private/club (access by pecuniary or non-pecuniary membership)	Gym, country club
Schools (membership or permission required)	Kindergarten, elementary, middle, high, private, and public school, college
Places of worship (free but selective/partial access)	Church, temples, synagogues, religious centres

socializing fabric of the locale, particularly for households with children. Sports amenities, such as the San Marino High School swimming pool, are open to use by residents, the sport fields at Huntington Middle School are regularly used for Little League games and for the annual fundraising by the PTAs, and its classrooms are used on Saturdays for the weekend Chinese School. See table 6.3 for the list of uses that are included in each category of gathering space and figures 6.2 (San Marino), 6.3 (Central Long Beach) and 6.4 (Mid-Wilshire) for the maps.

In the interviews, the residents were asked to indicate the places in their neighbourhoods where they were likely to experience intercultural contact. Their responses, categorized according to the access of gathering places in table 6.4, showed few places within the locales where neighbours could mingle and meet one another. This finding cross-validated with the routine maps of residents in figures 3.4, 3.8, 5.8, and 5.9, which illustrated that there were very few places for different ethnicities and nationalities to gather and engage, apart from the public neighbourhood libraries and parks. Comparing across the three locales, San Marino has more places for potential intercultural contact as compared to the other two locales.

Neighbourhood Parks

Parks and plazas host a variety of social learning opportunities. In fact, Frederick Law Olmstead, famous for designing Central Park in New York City, envisioned parks as "places where races could mix – off the plantation, in the city" (Sennett 2018, 44). They are "open-minded spaces" that Walzer (1995, 321) described as

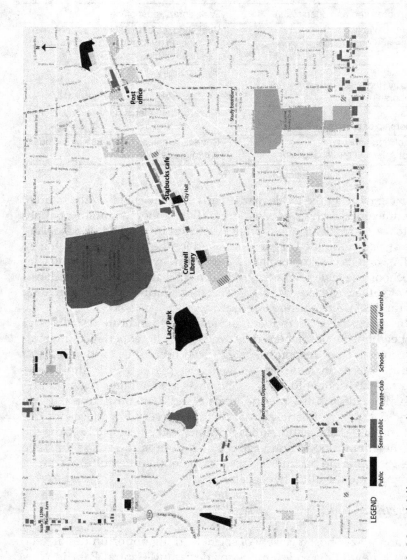

Figure 6.2. Map of different types of gathering places in San Marino. Map is not to scale.

Prepared by Haoyu Zhao and author jointly.

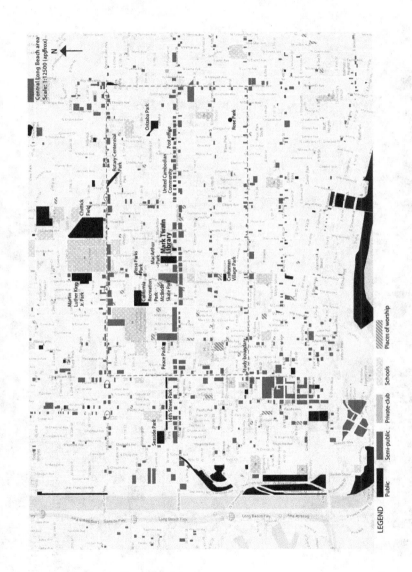

Figure 6.3. Map of different types of gathering places in Central Long Beach. Map is not to scale.

Prepared by Haoyu Zhao and author jointly.

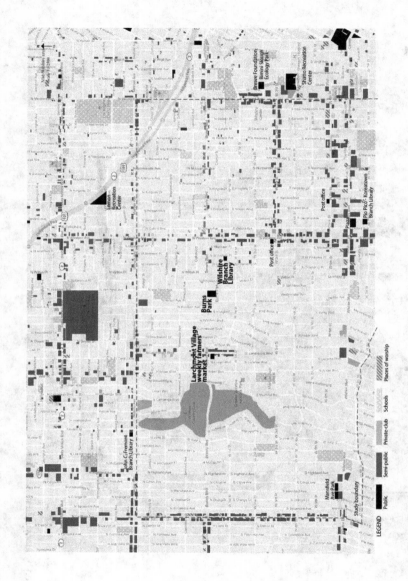

Figure 6.4. Map of different types of gathering places in Mid-Wilshire. Map is not to scale.

Prepared by Haoyu Zhao and author jointly.

Table 6.4. Places and events where intercultural learning is likely to occur in each locale, according to participants

Access	San Marino contact zones	Central Long Beach contact zones	Mid-Wilshire contact zones
Public (free and open to all)	**Lacy Park** **Crowell Library** The recreation centre Sidewalks: when dog-walking or strolling The post office	**Mark Twain Library** Sidewalks: when strolling or skateboarding Pickup area at the school	**Burns Park** **Wilshire Branch Library** The post office Larchmont Village weekly farmers' market
Semi-public (paid use and open to all)	**Starbucks cafe at San Marino Avenue**	Restaurants Grocery shops	Restaurants Shops
Private/club (access by pecuniary or non-pecuniary membership)	**Parent-Teacher Association (PTA) activities** **Chinese Club** **City Club** Rotary Club USC Alumni Club Homes of friends and relatives Workplaces San Marino High School swimming pool and field Churches	Homes of friends and relatives Churches	Korean Youth Association Homes of friends Tennis club
Events and festivals (usually open to all)	**Fourth of July picnic and fireworks** **Mid-Autumn Fundraising Festival** Christmas on the Drive Arbor Day Annual Pancake Festival, Hauntington Breakfast, Little League Baseball, Neighbourhood Watch meetings, block parties	**Martin Luther King Jr. Parade** Cambodian New Year Cinco de Mayo Neighbourhood meetings Neighbourhood clean-up Halloween Party Dance Fest	National Night-Out

Note: Local places that were most frequently cited by participants in **bold**

Designed for a variety of uses, including foreseen and unforeseeable uses, and used by citizens who do different things and are prepared to tolerate, even take an interest, in things they don't do. When we enter this sort of space, we are characteristically prepared to loiter.

Findings about the use of public parks in Los Angeles's multi-ethnic neighbourhoods indicate that the actual usage patterns of parks deviate from the intention. There was little mixing among social and cultural groups; instead, they coexisted in parks, carving out territories for their own activities (Loukaitou-Sideris 1995). The only space in parks where there was active mixing among different groups was the playground – among children and parents.

I observed similar patterns of use in the public parks in the three locales, which was that groups did not mingle at most areas of the parks, with the exception of the playgrounds. Burns Park, a small neighbourhood park in Mid-Wilshire with a popular playground, is an example par excellence of a social mixing. The playground is a spectacular multicultural as well as intercultural zone. On a typical afternoon, I saw groups of Spanish-speaking caretakers gathering at a shaded spot for a picnic while the White toddlers under their care played with each other; a Korean mom watching her son play while talking to the friend who had joined her in the park that afternoon; a Hispanic mom helping her child on the swing; an elderly Filipino nanny playing with an Asian girl under her care; a White nanny encouraging the little boy under her charge to climb another step up the slide ladder. In addition, children of different ages and different ethnicities played and socialized, creating opportunities for their caretakers who lived in different parts of Mid-Wilshire to interact. The fleeting "world of strangers" in Mid-Wilshire momentarily became a convivial world of familiar strangers. See figure 6.5 for photos taken in Burns Park.

Playing a soccer or basketball game in the public parks is an avenue for contact among the adults and teenagers from different social and cultural groups. According to those who had grown up in Central Long Beach such as Joshua Hernandez, Marteese Owens, and John Turner, there had been spontaneous but rare occasions during the entire length of their teenage years when Latino and African American teams played against each other in a friendly soccer or basketball competition. Whenever there was an opportunity to play together, the intercultural learning opportunities were immense. Participants reported being exposed to different styles of play, communication modes, and languages, and these micro-experiences had creatively challenged them to rethink group differences and similarities. Michael So, a Korean international student, described how these intercultural moments, gained through

Figure 6.5. Photographs of the different kinds of activities at Burns Park in Mid-Wilshire. (*top left*) Graduation party of a local high school, (*top right*) children building sandcastles, (*bottom left*) Spanish-speaking nannies picnicking near the playground, (*bottom right*) children enjoying popsicles after playing in the hot afternoon sun.

Taken by author in 2011 and 2012.

playing a game of basketball with strangers of a different ethnic group at the local park in Mid-Wilshire, made him feel more at ease when he encountered people from a different group during his daily activities.

> Even if it was brief, I like to talk to others, and I feel better about the other race when that happens. They are humans also. They like talking like me. They are good, they are nice. With more English I can get along with them.

As shown, public parks offer opportunities (though rare) that are empowering for social interaction with individuals of other social and cultural groups. For Yumi Lee, public parks were good spaces because they allowed for lingering, and lingering increased the occasions for conversations between strangers. As a new Korean immigrant in Mid-Wilshire, Yumi often felt insecure about speaking in English. But the leisurely atmosphere of Burns Park offered her a sense of calmness and safety that lessened the pressure to speak quickly, which could cause embarrassing mistakes in conversation. Going to the park helped her to communicate with strangers more easily and being able to do so in English was empowering for her. It made her feel socially accepted and gave her a greater sense of belonging in Mid-Wilshire.

Public Libraries

Public libraries are valued for their openness as a space that attracts a diversity of uses and users (Fincher and Iveson 2008). A distinctive characteristic of the public library, according to the residents I interviewed, was its safe and welcoming environment. Compared to a public park, which is usually unsupervised, a public library, with its twenty-four-hour security, is a comfortable space in which one can meet people safely and have regular interaction with others. Counter staff and patrons recognize each other over time and regularly talk to one another about their day.

In San Marino, parents viewed the Crowell Library as a safe space where their children could spend after-school hours before picking them up at the end of the workday. In Mid-Wilshire, visitors at the Wilshire Branch Library told me that the place was a friendly spot where complete strangers became familiar strangers over time, and this familiarity had helped to create an atmosphere of social trust. For example, people are willing to keep a watchful eye on bags or computers for the next person when he or she needs to step out to use the restroom or take a call. In Central Long Beach, visitors to Mark Twain Library felt comfortable in its casual but organized setting, in which users observed the proper etiquette, making the library a predictable and conducive environment, compared to the outdoor public areas. See figure 6.6. Marteese Owens, who prefers the library to the public park, has this to say:

> You've got tables, books, you've got a building and nice air-conditioning ...
> The police will come here fast ... You don't want to be open all the time,
> you got to be watching your back [like in an outdoor park].

The library is a place of refuge for individuals who live in locales with more street crime. The social and physical ease that users feel in a safe environment also enables them to become more open and willing to interact in public.

Figure 6.6. Photographs of the popular Mark Twain Library in Central Long Beach.

Taken by author in 2014.

The participants consistently referred to the neighbourhood library as a "neutral" space, one that was "not biased" space. It was highly valued for its open and free access to resources for a diversity of users. In locales of diversity where assertion of ethnic territoriality is prevalent, a place of neutrality provides respite that enables interaction among users on equal footing – an important criterion for productive intergroup relations according to social psychologist Allport ([1954] 1979) – where they could "mutually negotiate their common status as library users in the moments of their encounters" (Fincher and Iveson 2008, 188). Several immigrant participants said that witnessing how a diversity of individuals could respectfully interact with each other in a public space was empowering for them because it indicated that they themselves could also be included in the public life in the United States. Matthew Cruz, a new Filipino immigrant living in Mid-Wilshire, explained:

> It is important because you get to talk to each other, even if it is a brief time that you get to talk to each other. If you have a question and they answer you, that's a big thing, because you get to talk to people of another race or ethnicity with no problem.

The public library is a gateway to knowledge, has good infrastructure/resources for learning, and is an information centre for both residents and visitors. The public libraries in the three locales are also community spaces. Rooms in the libraries are available for reservation by neighbourhood groups for meetings, talks, and programs that cater to different age groups. However, in their current configurations of being predominantly places for quiet study, the potential of these three public libraries for encouraging spontaneous and organized social learning and intercultural engagement is limited. Incorporating "decompression spaces" like courtyard spaces with cafes where patrons could feel free to chat over some food and drinks within the grounds of the library would expand their utility into places where a wider variety of social interaction can occur.

Public Events and Festivals

Annual events and festivals in the locales draw diverse crowds, but for many residents, they have done little to improve intergroup relations. In the case of Central Long Beach, community organizers spoke highly of the annual Martin Luther King Jr. Parade (MLK Parade) as a social platform for intercultural learning where multi-ethnic civil rights are celebrated in public. Members from multiple social and cultural groups, representing their causes and interests, walk with props in their costumes along Martin Luther King Jr. Ave., passing through the entire Central Long Beach.

Their walk culminates at the Martin Luther King Jr. Park for a festival of food and celebration. See figure 6.7 for photos of the event.

My interviews with the residents revealed that events such as the MLK Parade, Cinco de Mayo, or the Cambodian Cultural Festival celebrated annually in Central Long Beach were not deemed by residents as intercultural platforms, but a celebration of one group's culture only. The residents felt that the events represented the separate interest of each major cultural group in the locale – Black American, Hispanic, and Cambodian – rather than a celebration of the diversity that they live within. While these annual events are intended to create publicity for multi-ethnic rights and greater visibility of the rich cultural diversity within the city, they do not substantially disrupt the status quo of the social exclusions experienced daily. They fail to offer new opportunities to empower and challenge residents to engage differently with one another. In the celebratory fanfare that accompanies these events, existing intergroup, dualistic tensions tend to be glossed over and remain unaddressed. Thus, while festivals are framed as the antithesis of everyday life, holding the promise to temporarily disrupt the status quo, they "also have the potential to support dominant power arrangements and inequalities," leaving them unchanged and entrenched (Fincher and Iveson 2008, 174).

The annual Mid-Autumn Festival organized by the Chinese Club of San Marino to fundraise for the local public schools faces similar challenges to provide an intercultural engagement platform that avoids further entrenching the dualistic tensions between groups. The Chinese participants spoke with pride about how the event, with a well-choreographed program that included traditional dance and cultural performances, offered their White American neighbours a good cross-cultural experience. The White American participants shared that while the cultural performances were educational, they did not help them learn about how to engage their Chinese neighbours. Cultural festivals tend to be unidirectional rather than interaction-based or intercultural. While these cultural events and festivals are well-attended, they have been ineffective in breaking the mode of co-presence in diversity. For example, the annual Fourth of July picnic and fireworks at Lacy Park in San Marino was for Jonathan Lin, a resident in his thirties originally from Taiwan who had lived in the city since his teenage years, an extended social club date. Jonathan explained that neighbourhood events often created extra opportunities for people who already knew each other to gather in their comfort zones rather than offering openings for strangers to break the ice and initiate new relationships. For him, these repetitive annual events and festivals were more effective in reconnecting and sealing existing relationships by providing a habitual and familiar platform for the process.

Figure 6.7. Photographs of Martin Luther King Jr. Parade along Martin Luther King Jr. Ave. in Central Long Beach.

Taken by author in 2012.

I asked the residents in the three locales for their thoughts on how their neighbourhoods could become better spaces for building relations among neighbours of different ethnicities and nationalities. Across the three locales, the suggestions were for more "third places" that are public and semi-public in accessibility, in addition to more interactive intercultural programs and events that could increase better access to information about cultures.

San Marino's residents suggested that a weekly local farmers' market (a "third place") could be a space to meet and mingle with neighbours, instead of just having to rely on social club meetings and chance encounters in the Starbucks cafe. In Mid-Wilshire, where existing gathering places are ethnically or religiously oriented, the residents suggested more organized activities accessible to the public, such as art festivals and sport events in the public parks, movies in parking lots, and a corner cafe attached to the Wilshire Branch Library. In Central Long Beach, residents had few places besides the public library that were accessible to all groups to intermingle. In addition, many residents underscored the importance of ensuring safety in the public places. Marteese Owens, whom we met in the previous chapters sharing his difficulties in building relations with his neighbours, felt that there was an acute need for a parochial, neutral, and safe place to encourage casual but sustained intergroup learning that could assuage the entrenched prejudice between African American and Latino youths. Marteese described the ideal gathering place:

> Maybe if you have some kind of program that brings people together with a good talking atmosphere, like the library kind of atmosphere. Everybody sits at the table and people talk of what they are doing. "I like music." "Want to play chess?" It could be done in the library. You got to have somebody who is promoting it, and actually start off that kind of program. Instead of the library closing, it turns over and then becomes – "We have something at the library, a get-together." A video-game night or something like that. It can be anything. You've got tables, books, you've got a building and nice air-conditioning.

Taking these sentiments into account within the context of the residents' responses, the challenge is not to stop holding neighbourhood festivals and events but to redesign them to fulfil their promise as a social platform to kick-start and develop new connections. As Fincher and Iveson (2008, 183) succinctly described,

> Rather, the point is that a good festival will create situations which make it easier for participants to step out of their conventional stances towards each other, enabling fleeting moments of encounter based on their shared status as participants in a festival.

Instead of focusing only on the celebration of cultural holidays or historical figures, events and festivals organized for intercultural exchange need to sensitively evince the difficult questions about diversity, such as prevalence of stereotypes, prejudice, and discrimination through the event. In addition, the momentum of intercultural learning sparked off at festivals needs to be sustained by programs such as regular activities and sustained conversations that would further develop these new connections into something more.

Concluding Thoughts: Public Realm of Diversity

These findings suggest three important and interrelated aspects of the public realm in diverse locales:

First, a public realm that enables collective life becomes highly desirable among a diverse group of inhabitants in places where fragmentary processes are at work. Practical and symbolic access to collective life in the public realm like at the public library acts as a centripetal force to counteract the centrifugal forces of social fragmentation experienced in the locales.

Second, the quality of intergroup contact in the public realm has an effect on the trajectory of social relations in diversity. Each experience of intergroup contact is embedded with tensions that can have dialogical or dualistic outcomes. In other words, a high-quality interpersonal contact in the public realm unleashes a dialogical learning process like Michael So's experience of playing basketball with youths unlike himself that can help to counter a misunderstanding or stereotype, and even form bonds of local belonging; conversely, low-quality encounters like Marteese's experience of the police breaking up multi-ethnic gatherings have the propensity to reinforce prejudices as these encounters can silently validate and reify mental categories.

Third, adroit design and programming of public environments in diverse locales is required due to the contradictions inherent in a diverse public realm to produce both dialogical as well as dualistic tensions, which could advance understanding and entrench misunderstanding respectively. A well-designed public place, like Burns Park in Mid-Wilshire where multiple social and cultural groups feel comfortable and safe to intermingle, is rarely found. In the following chapter, I will present an analysis of the socio-spatial qualities of diverse public environments and discuss the place design attributes that are conducive to intercultural learning and understanding.

7 Designing for Collective Intercultural Life

Let me begin this chapter with an excerpt of an interview with Lydia Li, a first-generation Chinese American resident of San Marino, about the relevance of public space for intercultural learning in diverse locales. A resident of the city since the 1990s, Lydia had participated in many civic activities and had been publicly active in different social circles within the city.

> MYSELF: How important are encounters in public spaces with people of another ethnicity?
>
> LYDIA: Public spaces don't really matter. You go there because you have a goal, and you are not there to meet people. People are brought together because of similarities in culture, in interests, in whatever. You can't force people.
>
> MYSELF: Do you think that how you are treated in public spaces, and how you see other people being treated, how you feel in a public space, the type of diversity there or the lack of it, makes a difference to removing some of the barriers that make relations more positive?
>
> LYDIA: In public spaces, there is much diversity. But you have no information about the people around you. Therefore, it is much easier for you to make conclusions based on immediate assessments, which oftentimes are not necessarily justified. For instance, if you go to the library and you see all these Chinese kids running around with no supervision, what would be your immediate reaction? "Those Chinese people!" You don't know about their backgrounds, but you see many Chinese there. The Americans pick up their kids and drop them off, or they are with their kids in the library. So, I think that public spaces are more detrimental to than helpful for better understanding [of others] because there is no source of connectivity.
>
> MYSELF: What if there is no public space – would it help understanding?

LYDIA: No. Neutral. I just think public spaces worsen [understanding]. So by not having public spaces, you don't worsen it. It is what it is. That's how I see it. Sometimes when you see something that is negative, it basically builds on your negative perception. It almost validates it even though it is inappropriate to use that as a validation. But people do use that to validate their initial assessment of who that person is. Public spaces are actually not good spaces for that, unless you use the public space to have multicultural or intercultural activities. Public space alone without any purposeful activity is almost negative.

The conversation with Lydia highlighted the perils and potentialities of public space as mediums of collective life in diversity, and that co-presence in diversity is no guarantee for conviviality. In fact, sociocultural diversity left unchecked and unmanaged, could enable the opposite effect of prejudice to grow. The scepticism that Lydia expressed about unprogrammed public spaces as stumbling blocks of social and cultural misunderstanding is also underscored by geographer Ash Amin's (2002, 967–9) critique of public spaces in diverse locales:

Diversity is thought to be negotiated in the city's public spaces. The depressing reality, however, is that in contemporary life, urban public spaces are often territorialized by particular groups (and therefore steeped in surveillance) or they are spaces of transit with very little contact between strangers. The city's public spaces are not natural servants of multicultural engagement ... In the hands of urban planners and designers, the public domain is all too easily reduced to improvements to public spaces, with modest achievement in race and ethnic relations ... The contact spaces of housing estates and urban public spaces, in the end, seem to fall short of inculcating interethnic understanding, because they are not structured as spaces of interdependence and habitual engagement.

To be sure, negotiating diversity in an unpredictable public realm is difficult. It does not help that public spaces in diverse locales are not designed to enable relationship building among different social and cultural groups. I see collective life as an intentional, interactive social life among groups, i.e., intercultural, rather than mere accommodation and toleration of differences. Collective life in diversity can be experienced through a variety of mediums, not least through the gathering of different groups in the same space, whether in physical space or cyber space. This book deals primarily with physical space and its potentialities for collective life in sociocultural diversity.

In this penultimate chapter, I want to discuss the extent to and ways in which public environments of diverse locales can *be planned and designed* to become productive intercultural spaces that facilitate dialogical tensions rather than dualistic tensions. The design of public environments matters because it shapes the conditions of social interaction between people and the possibilities for collective life in the city. Gathering the different threads of findings from the interviews, cognitive maps, and surveys with residents in Los Angeles, I will present an evaluation of the urban form of diverse public environments, discuss the scope of planning and designing interculturally, and identify the design qualities of public places that are productive in enabling and nurturing collective life in diverse locales.

Evaluating the Urban Form of Diverse Public Environments

Drawing on the eighty-seven cognitive maps and 140 interviews with residents, community organizers, and business operators, I have evaluated the legibility of the structure and identity of the diverse public environments by borrowing the criteria from Kevin Lynch's *Image of the City* ([1960] 1998). In his cognitive mapping study of users' wayfinding experience to understand the legibility of the urban form, he identified five common spatial elements of *districts, paths, nodes, edges*, and *landmarks* which shape the visual quality of the public environment of a city. In his later work, *Good City Form* (1981), Lynch wrote about the importance of the legibility of the urban form in enabling the development of a sense of a settlement, i.e., "the clarity with which it can be perceived and identified." He emphasized that for a heterogeneous society with multiple social and cultural groups, a clear sense of a city is particularly difficult to achieve, but it is certainly an important dimension of designing a good city where residents can develop a deep sense of place and belonging, in spite of different social and cultural values.

Given what we know about the configuration of the relational webs for territoriality along ethnic and national lines in multi-ethnic and multinational locales from the previous chapters, to what extent are the five spatial elements which emerged from studies of American cities in the 1950s, namely Boston, New Jersey, and Los Angeles, sufficient in explaining the legibility of the diverse public environment of a twenty-first-century metropolis? What new insights can we glean from how users navigate diversity spatially in contemporary Los Angeles that may offer valuable insights for the planning and designing of public environments that enable productive collective life in diversity?

Paths are the channels along which the observer customarily, occasionally, or potentially moves. They may be streets, walkways, transit lines, canals, railroads. (Lynch [1960] 1998, 47)

In diverse locales, the major roads were conceived as significant physical and social boundaries. For example, in San Marino, Huntington Drive – the main east–west arterial road – separates the wealthier residents and bigger houses in the north from the less wealthy residents and smaller homes in the south. In Mid-Wilshire, Western Avenue separates the low-income residents, who live in rental apartments, from their high-income neighbours, who own large houses and condominiums. In Central Long Beach, Cherry Avenue delineates the better and safer neighbourhoods in the east from the poorer and unsafe ones in the west. The major roads, with heavy traffic like that of Cherry Avenue and Western Avenue, also limit the physical and social connections between neighbourhoods across the roads. In the low-income areas of Central Long Beach and Mid-Wilshire, gangs also make use of major roads to demarcate their territories. In fact, when residents spoke about ethnic territories, they often referred to the streets where certain groups are concentrated or have more influence. Thus, paths in diverse locales, particularly in an automobile-centric city, are important socio-spatial elements as they are material and symbolic boundary markers.

Districts are the medium-to-large sections of the city, conceived of as having two-dimensional extent, which the observer mentally enters "inside of," and which are recognizable as having some common, identifying character. Always identifiable from the inside, they are also used for exterior reference if visible from the outside. (Lynch [1960] 1998, 47)

Residents of Central Long Beach and San Marino conceived their respective locales as having strong common boundaries and bearing the marks of a bona fide district. Due to their unique social, economic, and cultural conditions, these two locales stand out from their surroundings. Central Long Beach was conceived by its residents as demographically and environmentally distinctive as a district with poverty, high population density, and ethnic diversity. In contrast to the rest of the City of Long Beach, where the physical landscape suggests more care, control, and homogeneity, the buildings in this locale appear worn and old, the signages of shops here are often not in English, and there are many more neglected vacant lots. San Marino stands apart from its surroundings as a district of wealthy residents with big homes, well-maintained lawns, and no ethnic food smells; it is ethnically more diverse (due to its substantial White population) than the neighbouring cities with large Asian enclaves. In the case of Mid-Wilshire, the multiplicity of political, cultural, and socio-economic affiliations produces *multiple subcultural districts* in linear, horizontal, and vertical forms. For example, Koreatown, Little Bangladesh, Hancock Park, and the Salvadoran district are horizontal and linear subcultural districts recognized by both residents and visitors. Vertical subcultural districts are harder to discern for the passer-by, but long-time residents with keen social and spatial intelligence can quite easily identify which street and which building has an ethnic concentration.

From the above, we can see that a common characteristic in these landscapes of diversity is the *presence of enclaves* of ethnicity, nationality, language, wealth, and poverty. In Mid-Wilshire and Central Long Beach, according to residents, the enclaves were formed over time as more people belonging to the same ethnicity or nationality moved in, displacing other residents of a different ethnicity or nationality, and the services that supported the group also accumulated as the group grew. Linguistic isolation is not uncommon in these locales. In contrast, the residential enclaving patterns are not as visible and as spatially concentrated in San Marino. Instead, enclaves in San Marino are social ones formed through socialization patterns and club memberships.

The element of district is very strong in diverse locales as sociocultural differences are manifested spatially, constantly differentiating the insider from the outsider and drawing legible boundaries that are detectable via the sense of sight, sound, and even smell.

> **Edges** are the linear elements not used or considered as paths by the observer. They are the boundaries between two phases, linear breaks in continuity: shores, railroad cuts, edges of development, walls. (Lynch [1960] 1998, 47)

Locales of diversity have *multiple edges* that are usually not physical or visual but social and symbolic. In some instances, these edges can also have physical forms. In Mid-Wilshire, myriad intersecting lines of differentiation – ethnicity, nationality, income, lifestyle, and values – have produced a fragmented social space of *irregular edges*. The irregular edges create border conditions that have varying levels of permeability (physical and social) that could catalyse friction, collision, and dualistic and dialogical tensions. The degree of social permeability in a diverse locale is highly dependent on an individual's affiliations, resources, and intercultural capabilities to negotiate the border dynamics. In the case of Mid-Wilshire, there are very few gated areas, so its edges are largely physically porous. However, its social permeability is not high because of the prevalence of "ethnic bubbles" that made building intergroup relations difficult, particularly in the eastern neighbourhoods.

In Central Long Beach, crime and poverty have commingled with ethnic gang activities for many decades, producing the edges around the neighbourhood that are more like a fortress wall, keeping those on the outside away and trapping the residents within them. However, the porosity of the edges varies depending on the time of the day. For example, Twentieth Street, which is used as a routine path by children walking to school in the day, has been identified by residents as a place of street violence at night between rival gangs, who are fighting to claim new territories.

San Marino's high property values have boosted the property prices of its immediate surroundings, which means that the houses in the adjacent cities along the municipal boundary tend to look like the ones in San Marino, hence masking the physical edges of the city. As such, in this city, edges are more symbolic and social, rather than physical. From the interviews, it became apparent to me that club memberships (e.g., the parent-teacher associations [PTAs], City Club of San Marino, Rotary Club, and university alumni clubs) organize the patterns of social life in the city and play a major role in creating new inter-group affiliations, as well as accentuating differences between members. The differences in language, migration experience, ethnicity, nationality, life stage, and wealth among the residents are symbolic and social edges that differentiate and separate people further into cultural subgroups.

Edges are pervasive and highly relevant elements in diverse locales. They are in fact integral parts of the urbanism of diversity. Their presence and function go beyond the physicality of Lynch's original intent, taking on social and symbolic meanings that signal either opportunities for intercultural mingling, or a resistance to forms of collective life in diversity.

> **Nodes** are points, the strategic spots in a city into which an observer can enter, and which are the intensive foci to and from which he is traveling. (Lynch [1960] 1998, 47)

Based on the residents' cognitive maps, there were very few conventional nodes per Lynch's original formulation in these diverse locales. Unlike conventional nodes, which are usually located at major street intersections, nodes in locales of diversity tend to emerge from places of ethnic gathering. In Central Long Beach and Mid-Wilshire, the large availability of ethnic shops and services creates *ethnic/nationality nodes* amid the diverse demography of the areas. These ethnic nodes make up a huge portion of nodes. The remaining nodes are the few civic amenities such as libraries and neighbourhood parks where paths inter-sect and inter-ethnic socializing may take place. In San Marino, fund-raising events organized by schools and social clubs form the major *socializing nodes* in this city whose residents tend to be reserved and private. Most of the socializing nodes are thus temporal and invisible to the public realm. The only notable node identified by many residents is the Starbucks cafe opposite the city hall.

Therefore, the element of nodes remains relevant as a dimension of diverse locales, but their forms are less integrated with the physical, infrastructural layout of the city. Instead, nodes in diverse locale are organized according to social activities and territoriality defined by groups. In view of this, mapping the location of nodes can reveal the contours of collective life in the public environment of a diverse locale, which is helpful to planning the land use and public places in diversity.

> **Landmarks** are another type of point-reference, but in this case the observer does not enter within them, they are external. They are usually a rather simply defined physical object: building, sign, store, or mountain. Their use involved the singling out of one element from a host of possibilities. (Lynch [1960] 1998, 48)

In locales of diversity, signages in different languages and non-Roman alphabet scripts on shop fronts, billboards, and civic organizations are *soft landmarks*, in contrast to the classic brick-and-mortar architecture of hard landmarks. In Central Long Beach, signboards in Khmer script and romanized Chinese words stood alongside those in Spanish and English. In Mid-Wilshire, Korean, Spanish, and English signs and bill-boards line the streets. In San Marino, business signs are usually in English with an occasional Chinese subscript. Another soft landmark in these landscapes is the presence of unique businesses not commonly found in mainstream American parts of LA. These include pawn and jewellery shops owned and operated by Cambodians in Central Long Beach, retail coffee, bakery, and business chains from South Korea in Mid-Wilshire, and Chinese-style after-school tutoring centres operating along Huntington Drive in San Marino.

In Los Angeles, residents in diverse locales do not actively struc-ture their everyday sense of place using landmarks. This is perhaps an outcome of the car-centric repetitive landscape of strip malls where distinctive architecture is sparse. Instead, road signages, unique archi-tectural facades, or gateway arches are helpful for visitors and outsiders as visual signals that the space they have entered is ethnically diverse or ethnically different. However, this is not to say that landmarks are unimportant. Wherever they are found, whether in the form of buildings, murals, or sculptures, landmarks in a diverse locale with multiple resident immigrant groups are usually politicized, symbolically marking ownership, asserting group identity, and legitimizing the spatial claim of one group over the others in the city.[1]

Overall, this study found that the elements of *districts* and *edges* are very dominant socio-spatial elements in the public environments of diverse locales. In addition, the element of *paths* is an important but secondary feature in the organizing of legibility and identity of diverse locales. As such, the findings of this study on diverse locales deviate somewhat from other cognitive mapping studies done on Los Angeles that indicate that *paths* are the predominant element structuring the sense and legibility of a locale.

For example, in Lynch's ([1960] 1998) study of Downtown Los Angeles, he found that in the images produced by residents, the city was an "undifferentiated matrix" of the street grid, there was a primary reliance on street signs for orientation and legibility (p. 33) and there were "no medium-sized districts" (p. 41). In another study by Banerjee and Baer (1984), in which 375 mental maps drawn by residents from different income neighbourhoods across Los Angeles were analysed to understand how the built environment shapes the well-being of residential areas, they found that residential areas in the city lacked well-defined districts. Instead, the elements of paths and nodes most frequently emerged in residents' cognitive maps of the neighbourhoods. Last, in Cheng's (2009 and 2013) study of the racial formation of San Gabriel Valley, thirty mental maps by residents were analysed. Like Banerjee and Baer (1984), Cheng highlighted that participants structured their images of geographical space by major roads and freeways which also function as "important dividing lines."

In comparison, this study illustrates the dominance of districts and subdistricts in diverse locales. Navigational legibility is not merely physical and reliant on road signs, but social, as residents share a heightened awareness of group territoriality in the public environment. Residents depend on a combination of their daily spatial practices, lived experiences, and mental impressions of group relations and locations to map their legibility. This finding underscores that relational dynamics play a critical role in shaping the legibility and identity of the public environments in a diverse locale as compared to a socioculturally homogeneous locale. In fact, the social complexity of its physical public environment necessitates an active inclusion of residents in the planning and designing of public places to ensure equitable access and productive use of public places for collective life.[2]

Interculturalism in Urban Planning and Design Practice?

Handling "Diversity"

In the interviews with urban planners and park planners, they spoke keenly about how their locales had adapted to their diversifying demographies. They gave the example that community engagement is now done with real-time translation services and that the process of decision-making in planning makes a deliberate attempt to include more stakeholders. The planners interviewed viewed "diversity" as a descriptive label referring to "a mix of culture and environment," or one that is simply "a collection of race, ethnicity, income businesses, residents, densities." During the interviews, the planners of the three locales were very mindful to avoid "calling out" certain groups and cultural references in their discussions of diversity. Instead, the planners relied on the abstract meaning of an incorporation or "mix" to explain the patterns of diversity, even though the realities on the ground reflected a sorted rather than mixed form of diversity, as evident in the discussions of previous chapters. Overall, the planners exhibited a heightened political sensibility and some empathetic sensitivity about diversity as a phenomenon important for the planning of a cosmopolis, which Sandercock (2003) has underscored. However, they did not question established design norms or planning mores.

When the cognitive maps of the planners were compared against those of the residents, it was apparent that the planners' conceived space of the locales was more simplified than that of the residents. The planners' maps did not have much information about the social geography of the locale, such as how ethnic groups were distributed over space or how they used or moved through it. In addition, the planners had mapped the urban space of the locales along contours of economic opportunity – more functional than relational, more discrete than nuanced. This was the case in Central Long Beach and Mid-Wilshire, where the planners demonstrated a notable bent towards framing ethnic diversity as a functional leverage for economic development in the low-income neighbourhoods. As one planner said,

> [Having multiple ethnicities] is a status quo. We, as a city, are very used to that fact. We don't see, in terms of planning and planning outcomes, that ethnicity matters that much. It is really about how we attract investment and how to make [the area] look as nice as we can make it to look.

Normative Intent, Technical Means

As a practice, urban planning is guided by a normative intent for universal betterment and relies on positivistic knowledge and technical tools to execute its vision (Friedmann 1987).[3] For example, zoning, a basic technical and rationalizing tool to plan and allocate land uses by systematically separating uses according to functions, was born out of a desire to preserve and protect uses valued by society. The irony is that once these technical rationalizing tools are developed, urban space is framed in the language of land uses and functions, rather than a medium of social relations.[4]

When the urban planners and park planners of the three locales were asked about the relevance of intercultural understanding for the planning of the locales, the former were pragmatic in their responses. In San Marino, the scope of intercultural understanding was less about relations between residents than it was between the planners and the residents. For them, intercultural learning and understanding occurred when interaction revealed the cultural contexts behind the remodelling design preferences of residents that were not immediately aligned with the planning parameters in the city. In Central Long Beach and Mid-Wilshire, planners saw intercultural understanding as enabling a process of social interaction between residents. However, the planners quickly added that this is not what they were tasked to do.

They saw urban planning and design as a vehicle to deliver the hardware or physical infrastructural space, rather than the software or social development of intercultural dialogue and understanding in these spaces. They explained that in urban planning, the planners care about safety in public spaces, which is a multi-dimensional issue for residents. But the scope of intervention for planners is mainly physical and infrastructural, through ensuring that there is adequate street lighting, security cameras, clear paths of travel, and graffiti resistant surfaces. Similarly, for the park planners, they emphasized that their design consciousness is inclusive, rather than intercultural per se. While planning and designing parks for a diversity of users is a primary goal for them, diverse locales oftentimes have complex and conflicting user needs so that as designers of public places, their scope is to ensure general physical accessibility and safety in the layout of a public place, rather than to facilitate social interaction among groups. This quote from an urban planner in one of the locales sums up the scope for interculturalism in urban planning and design:

> I think the main thing about planning and zoning is that they deal with the built forms, intensities, scales. It is not directly dealing with those

components [*referring to intercultural learning and understanding*]. I think it is a secondary impact or consideration. Obviously, we are trying to focus on a pedestrian-friendly neighbourhood that creates an environment where you might have a higher degree of this mutual understanding and interaction happening. But that is a secondary impact. We are not social engineers, although other people might say we are. At the end of the day, when a project comes in, we are looking at the zoning code, we are looking at design. Not that it is not important. It is just not 'planner-y.'

As a result, the substantive outcome delivered by planning inadvertently becomes less about how people relate in and through space; rather, it becomes more about how land uses and functions of the cities relate to each other. These planning tools can quickly become the raison d'être of planning practice, offering planners the distance from a normative and value-based practice that is vulnerable to the fragmentary stresses produced by different and diverging sets of values in diversifying cities.

Universalizing Needs and the Equity of Difference?

In addition, planners' goal of meeting universal needs (defined by basic demographic characteristics like age and life stage) with the objective parameters of development offers a less contested ground upon which to represent the public interest. Introducing interculturalism into planning challenges this well-defined paradigm because the goal to enable an equity of difference is inherently difficult to translate into objective, universal measures. However, with greater societal schisms in the second decade of the twenty-first century arising from a complex gamut of political, economic, and racial inequalities, the call to incorporate diversity equitably and effectively has become even more salient.

Planning scholar Susan Fainstein (2005 and 2010) has argued that incorporating social diversity as a principal goal of planning may not offer equitable outcomes, while other scholars like Burayidi (2015) and Qadeer (2016) have presented arguments that a multicultural accommodation of cultural differences through adapted planning parameters would offer greater equity to social and cultural groups in diversity than simply focusing on improving interaction via interculturalism. In *Cosmopolis II: Mongrel Cities in the 21st Century* (2003), Sandercock proposed that the making of a cosmopolis requires new sensibilities in planning practice that include growing intercultural competencies and being ready to address conflicts through a therapeutic approach that allows for emotional engagement of intergroup relations. The normative ideal of a cosmopolis is idealized in the equitable access to spaces of free intermingling of differences without exclusion, per Iris Marion Young (1990).

For others, achieving racial equity amid diversity in cities requires a more radical change that includes overturning urban planning's legacy of supporting White privilege (e.g., Goetz, Williams, and Damiano 2020, Solis 2020, Williams 2020). This goes beyond achieving socio-economic equity by adopting an "anti-subordination" approach in planning as proposed by Steil and Delgado (2019, 42), which supports "policies that identify and address historical inequalities and relations of domination along the lines of sex, sexuality, race, ethnicity, national origin, and other ascriptive characteristics."[5] According to Steil and Delgado (2019), who built on Fainstein's (2010) principles of diversity, one important way to incorporate diversity fairly in the public environments of cities is to remove the disparity among groups to access public spaces and to co-produce these common grounds with disadvantaged groups. Public spaces are normatively critical to the practice of everyday life and symbolic of societal equity and inclusion.

Incorporating a form of interculturalism that also aims for equity in planning requires planners to not only deliberately include in their practice fact-based, objective, universal parameters, but to also negotiate the vagaries that come with the subjectivity of different cultures. This is a challenge that is difficult to accept, especially when planning is deemed "acultural," as the planner of one locale told me. To borrow Umemoto and Zambonelli's (2012, 204) phrase, planners need to "facilitate deliberation among people who share different 'ways of knowing.'" In other words, planners need to be prepared to unpack public interest into the interests of multiple social and cultural groups that have different sets of values and perspectives that sometimes agree, and at other times, conflict with one another. They have to commit to doing this while guarding against an outcome where everything goes, and social inequalities become further entrenched. It is a tall order. Fully engaging with interculturalism in planning can perhaps be seen like the opening of a Pandora's box of discontent, injustice, and contestation amid the promises of meaningful collective life.

Sensitivity to Intergroup Tensions and Their Manifestations

Incorporating interculturalism also requires new knowledge and sensitivity to intergroup tensions and their manifestations in the spaces of the city. Writings by Bollens (2006), Gaffikin, Mceldowney, and Sterett (2010), and Gaffikin and Morrissey (2011) about post-conflict cities where intergroup differences have splintered societies violently such as Belfast, Nicosia, and Barcelona underscore the importance of sensitivity in planning to the tension lines between groups, especially in locating new neighbourhoods and marketplaces, and in creating physical and social connections through transportation planning. According to Bollens (2006, 67), the spatial organization of the city is a strategic social infrastructure that can enable contact opportunities in circumstances of divisions, differences, and diversity, through which abstract concepts of "democracy, fairness, and tolerance" are put into practice. Bollens (personal communication, 2009) emphasized that planners have to be cognizant of the extent of permeability of the boundaries between groups in sites of conflicting interests and of the different kinds of territories, otherwise reconciliatory efforts could easily be thwarted, becoming divisive actions instead.[6] As Amin (2010, 3) posited, "interventions in the urban unconscious (public spaces, physical infrastructure, public services, technological and built environment, visual and symbolic culture) have an important role to play in regulating social response to differences."

To borrow a metaphor from the electronics world, the above dilemmas point to a disjuncture between a reality that is *analogue* – continuous and fuzzy – and planning practices that rely heavily on *digital* approaches – binary and discrete. Planning well in diversity applies pressure on urban planning to confront the challenges thereof. It requires going beyond the first steps of recognizing diversity and adapting the planning process to include stakeholders for consensual decision-making. To secure substantive equitable and convivial planning outcomes amid social and cultural differences, planners should be intentional about facilitating intercultural relations through their design of public spaces and planning of land use in the city. This is particularly critical in locales of diversity where the harms and potentialities of social tensions are present. Writing about conviviality in cities, urban anthropologist Lisa Peattie (1998, 248) is of the view that urban planning plays a critical role in its co-production:

> In human happiness, creative activity and a sense of community count for at least as much and maybe more than material standard of living. Planning can enhance the possibility for conviviality ... Conviviality cannot be coerced, but it can be encouraged by the right rules, the right props, and the right places and spaces. These are in the domain of planning.

In the quote above, Peattie alludes to the importance of planners taking a mindful and light-touch approach when shaping the dynamics between physical space and social relations.[7] It is clear that conviviality is like a dance that requires planners and designers to understand and work with residents to co-design the choreography and be co-producers of the show.

Co-producing a Convivial Collective Life: Qualities of Intercultural Places

My view is that the concept of co-production is in agreement with interculturalism as a process of mutual learning through engaging in common activities and intentional dialogue between groups. Co-production of intercultural spaces calls for planning and designing with an understanding of users' intergroup dynamics and lines of tensions so that productive social interaction can be catalysed. This understanding can come from talking with and learning from users through interviews, surveys, and ethnography, as well as "brokering" among different users to bridge differences. The task of brokering for planners is described by Rios (2015, 358): "to emplace ourselves with others and continually discover how to speak (verbally and nonverbally) and act between and among different cultures with the goal of enabling greater exchange, understanding, and ultimately, respect."[8]

Designing for Intercultural Learning: Eight Place Qualities

In their respective empirical research studies, sociologist William H. Whyte (1980) and urban designer Jan Gehl ([1987] 2011) found that there is a correlation between the quality of interaction among strangers and the physical configurations of urban space, such as the availability of seating and food, the angle of the sun, visibility of the space, location, and more. Take for example the availability of seating. Seats invite and allow people to slow down, pause, and linger in a space, and hence offer the possibility of starting a conversation with other people.[9] According to Gehl ([1987] 2011, 159), a public space "should offer many different opportunities for sitting in order to give all user groups inspiration and opportunity to stay." Seating arrangements in public spaces should aim to be "socially comfortable," according to Whyte (1980, 28), such that choice could be given to users – whether it is "sitting up front, in back, to the side, in the sun, in the shade, in groups, off alone."

In Risbeth and Rogaly's (2018) study of the urban bench, they found that opportunities for people to sit and watch are a crucial means for residents to partake in the collective life of the neighbourhood. In their research about Gordon Square in Britain, they observed that the act of

Figure 7.1. The qualities of public places that encourage intercultural learning.

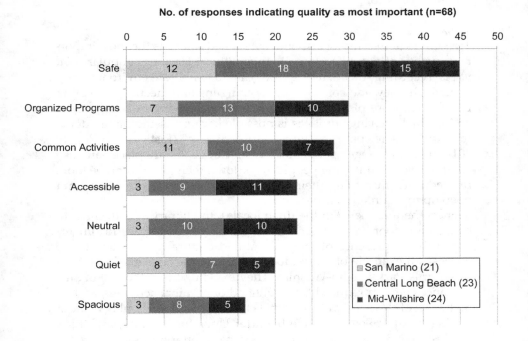

No. of responses indicating quality as most important (n=68)

watching the programs on the large public screen in the square was one made, by some people, against self-isolation. They found that public benches gave people the "right to pause," and they were a site of potentiality for social interaction and inclusion. The availability of public benches in Gordon Square gave elderly Nepalese (a minority population in Woolwich) a place to gather, providing them with public visibility and a sense of local belonging. Thus, the availability of seating boosts the likelihood of users of public space to be included in collective life and increases the ease of strangers to initiate convivial connections with each other.

With an eye towards the study of the co-production of intercultural spaces, I created a short survey to find out what the participants regarded as qualities of public spaces that are of paramount importance in initiating and developing intercultural relations in their respective diverse locales (see figure 7.1 for the responses). There was a total of sixty-eight participants who filled out the survey from the three locales, whom I will refer to as survey respondents in the following discussion.

1. SAFE

Two-thirds of the survey respondents ranked safety as the most important quality of a public place for intercultural learning. Popular *third places,* according to Oldenburg (1989), are usually safe environments – relaxing, homey, and comfortable for mingling and meeting with other people. They are places where people enjoy hanging out, and where an informal exchange of ideas is possible because of the camaraderie. In addition, safe spaces create a "notion of ease" (Risbeth and Rogaly 2018) that encourages people to stay and linger.[10] Safe places are also comfortable, encouraging people to let down their guards and be more relaxed, thus increasing their willingness to interact with others and open up to an unfamiliar person.

From the interviews in the three locales, the library was deemed to be a safe indoor environment and thus was a good space for lingering in the co-presence of other social and cultural groups across the locales. An example of an outdoor space that participants felt safe in was Burns Park in Mid-Wilshire. They explained that the use of surveillance cameras reduced the likelihood of criminal activities, and its high level of cleanliness ensured that children could safely play without getting cut by broken glass or other litter. Upon further analysis, I think the design of Burns Park played a major role in offering a sense of safety to its visitors. Sloped like a miniature amphitheatre, it is an open and airy environment with good visibility. The playground, located in the middle of the park, has a short fence around it and caretakers can easily keep an eye on the children while seated on the grass. Further, the safety in Burns Park is enhanced by the caretakers who form a natural community that watches out for every child, in a way that is akin to how urban streets are kept safe by shopkeepers and pedestrians who have their "eyes on the street," on the lookout for subversive activities (Jacobs [1961] 1989).

2. ORGANIZED PROGRAMS AND COMMON ACTIVITIES

Many respondents indicated that public spaces needed to have organized programs and common activities in order to encourage people to step out of their comfort zones and interact interculturally. Safety is necessary for ease but not sufficient for active engagement in border-crossing conversations or joint activities. In my interview with Robin Toma, Executive Director of the Los Angeles Human Relations Commission, and in a separate interview with Ray Regalado, a senior intergroup relations specialist in the same organization, they expressed a common view that there is a strong pull towards socializing within groups in Los Angeles. They think that there is a dire need to create multifaceted programs in schools, neighbourhoods, and workplaces to help people interact and engage with individuals from other social and cultural groups. According to Amin (2002, 970), by involving people from different groups in new common activities, the process "disrupts easy labelling of the stranger as enemy and initiates new attachments."

In planning and designing community spaces in locales of diversity, thought could be given to central urban spaces that could host programs and opportunities for common activities among youths and adults. For instance, municipalities could plan events such as soccer matches or basketball games among different cultural groups, which could be held in the parking lot of a well-located mall. They could also design flexible spaces in parks and libraries for regular activities, like community gardening paired with outdoor cooking lessons that teach American and international cuisine. These opportunities provide residents with common platforms to facilitate the making and development of new connections through joint participation within the realm of the neighbourhood.

3. ACCESSIBILITY AND NEUTRAL

Accessibility may be interpreted in two interrelated dimensions, physical and symbolic. The vitality of public spaces is dependent on its accessibility to a mix of users and uses.[11] Symbolic accessibility, i.e., whether one feels welcome to use the space, is crucial in diverse locales, where social tensions arising from territoriality are evident. How inclusive or exclusive a public space is is usually determined by who its users are and the patterns of use. Does a space host a diversity of users or is it divided up into territories and occupied by social and cultural groups, so that the public space is no longer a neutral ground?

According to Oldenburg (1989, 22), popular *third places* are characterized by spaces that have easy accessibility and an absence of power imbalances, i.e., where "individuals may come and go as they please, in which none are required to play host, and in which all feel at home and comfortable." A neutral public space invites different social and cultural groups to share its space. Neutrality enables diversity and the possibility of developing intercultural relations. About a third of the respondents ranked neutrality of public space as a very important quality of a public space for intercultural learning, particularly those participants in Central Long Beach and Mid-Wilshire. This response further supports the findings in previous chapters that show how social space in these two locales is fragmented by vertical and horizontal ethnic enclaves, income polarities, and gang territories.

Neutral spaces require intentionality in their design and implementation to keep them accessible, especially in diverse locales. With regards to this aspect, there is much to learn from the example of Siena, the Italian medieval city where warring families had periodically engaged in violence to stake out territories. Its city council, in a bid to quell the social disorder, selected a sloping ground that did not belong to anyone, and transformed it into the commons of the city. Piazza del Campo came into being as a programmed space to serve as the neutral ground for the negotiation of difference between warring families, through the friendly games and competitions held there. The Siena Palio, a famous horse race, is one such (historic) game that has continued up to modern times.

4. QUIET AND SPACIOUS

Spatial ergonomics has been found to affect the quality of interpersonal contact and the potentiality of human connection. Neither a noisy or very quiet street corner is conducive to lingering or conversation. Likewise, the proportions of a space affect its attractiveness as a place to hang out in. A small space might evoke claustrophobic feelings in some, and an overly large space might be overwhelming. Urban acoustic research has demonstrated that a good place for conversation should not have background noise that is beyond sixty decibels. In a space with a noise level beyond sixty decibels, people sit or stand a lot closer than may be socially comfortable. According to Gehl ([1987] 2011), a socially comfortable space is one that has a maximum distance of twenty-five metres (eighty-two feet), in order for people to see facial expressions, and should not exceed 100 metres (360 feet) for observing events.

While these spatial ergonomic qualities are important for the creation of enjoyable urban spaces, they are all the more important in environments of diversity to mitigate the cultural barriers and social anxiety that make spontaneous and productive interaction difficult. However, are spatial ergonomics alone sufficient to catalyse new intercultural connections and develop convivial relations?

Through the interviews with residents in the three locales, it became clear that socioculturally diverse places are milieus with multiple mores and expectations that make spontaneous interaction between strangers (especially with a person of a different ethnicity) hard to initiate. In such socially sensitive circumstances, icebreakers are helpful disrupters of the status quo and are social triggers that can facilitate identifications between strangers.[12] Icebreakers are essentially communication catalysts to nudge the exchange of information among strangers, so that new lines of connection can be established through found commonalities or middle grounds amid differences. The mechanism of the icebreaker is similar to the principle of "triangulation" that William Whyte (1980, 94) identified in *The Social Life of Small Urban Spaces*. Whyte defined triangulation as "a process by which some external stimulus provides a linkage between people and prompts strangers to talk to each other as though they were not." These external stimuli must not be overly planned and scripted, but they must be just enough to spark conversations and learning moments. Fincher and Iveson (2008) call this "light-touch programming." The interviews with the residents of the three locales in this study highlighted a few significant catalysts that were effective in activating productive face-to-face contact between familiar and complete strangers of different cultures in public spaces. They include the following:

5. FAMILIARITY

Familiarity can breed contempt but also congeniality. On one hand, familiarity with differences can create indifference to diversity because relations become routinized, presumptive, and categorical, resulting in the phenomenon of the familiar stranger per Milgram ([1972] 2010). On the other hand, familiarity is "a benign experience of repetition [that] can in and itself enhance positive affect," according to Zajonc (2001, 224) and thus, provides a basis upon which to initiate contact among strangers in diverse public settings. In Suzanne Hall's (2012) ethnographic study of a diverse street in London, she shows that familiarity can even affirm an individual's belonging to a place and provide him or her an enhanced competence to negotiate social life in diversity.

As part of the survey, respondents from the three Los Angeles locales were also asked to rank a list of places and events in their neighbourhoods based on which they thought would be good places for starting a conversation – one that would lead to intercultural learning and understanding – with someone of another ethnicity or culture. The respondents were asked to differentiate between the familiar stranger (defined in the survey as a person they visually recognize) and the complete stranger (defined as a person whom they are meeting for the first time). Figure 7.2 illustrates that visual familiarity is critical in improving the likelihood of residents of different ethnic groups initiating conversation with one another in every location in a diverse locale. Familiarity is developed through repeated encounters. It eases the anxiety of unpredictability of diverse locales. It also serves as an icebreaker for conversation and other forms of non-verbal interaction between strangers.

Figure 7.2. Places for conversation with familiar and complete strangers in the neighbourhood (N=61). Y-axis shows the cumulative score based on the inverse of the rankings.[13]

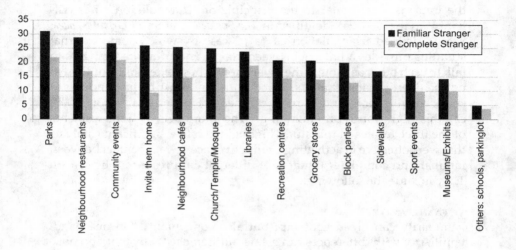

When seen in this light, brief encounters in the routine spaces of the neighbourhood such as the sidewalks or grocery shops are not insignificant. Repeated encounters of familiar strangers can pave the way for more durable, future intercultural relations.

The findings from figure 7.3 highlight that locales of diversity need a variety of public, semi-public, and open spaces, both indoor and outdoor, where familiar and complete strangers can have opportunities for frequent face-to-face meetings to kick-start and develop relationships.[14] They also indicate that certain venues are perceived by respondents to be more productive for establishing contact between familiar strangers who are from different social and cultural groups. According to figure 7.3, the survey findings illustrate that semi-public places like local restaurants and cafes are opportune for intercultural learning between familiar strangers. These *third places* (Oldenburg 1989) create an atmosphere of safety and openness to socialize and linger. In contrast, public parks, libraries, places of worship, museums, school, parking lots, and community events are spaces that allow new connections between complete strangers to form (more) easily. The best locations for familiar and complete strangers to intermingle, linger, and socialize are public parks and community events. They are the *cosmopolitan canopies* of diverse locales, where "people of differing racial and

Figure 7.3. The likelihood of conversation between familiar strangers in various locations/places. Y-axis refers to the ratio between the cumulative score of familiar stranger to complete stranger.[15]

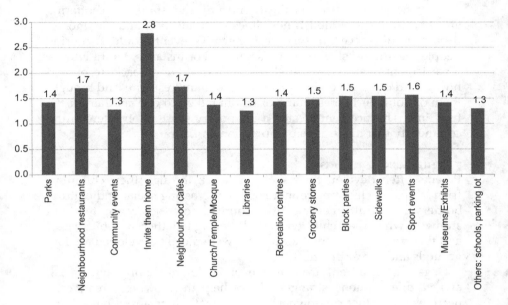

cultural types not only share space but seek out each other's presence" (Anderson 2011, 30).

6. THE UNEXPECTED AND CURIOSITY

An event that occurs unexpectedly in a public place could pique curiosity in those who encounter it and initiate contact between strangers both familiar and new.[16] A catalyst of learning and growth, curiosity is often found aplenty in children, especially about the unknown and unexpected. For example, a flash mob could evoke feelings of togetherness and energy in bystanders, or a street magician could create a communal sense of wonder and openness in unsuspecting participants. This sense of shared curiosity and experience with strangers can create new common dialogical grounds that are capable of triggering conversation between people from different social and cultural groups. The use of large, interactive screens in public spaces to engage community members in spontaneous discussion of topics of social interest via their personal mobile devices has also been effective in triggering an experience of public intimacy and culture (McQuire, Papastergiadis, and

Cubitt 2008). These "affects of togetherness" (Amin 2010), even though evanescent, are significant moments of collective life in diverse settings that are vulnerable to dualistic tensions.

The unexpected disrupts the humdrum and status quo, offering an opportunity to establish new lines of connection and commonality across differences. In this sense, negative events could also bring people together in shared commiseration. For instance, Laura Nunez, a participant interviewee who lives in Central Long Beach, recounted how a loud car crash on her street startled many residents and brought them out of their homes, many of whom talked to one another for the first time. This accident created a temporary sense of an intercultural community that made a deep impression on Laura.

7. CHILDREN AND DOGS

Parents often develop their social network through their children's friendships, as children are more adept at crossing social and cultural borders than adults. Participants recounted how they connected with strangers when their children started playing together in public parks, and that friendship with other parents were mostly developed at playgrounds and soccer practices.

Dogs, like children, are effective props for mediating and easing awkwardness among strangers. Dogs help their owners create connections with other dog-owners. As dog-walking is a routine activity, dog-owners regularly meet one another in the neighbourhood, like on the sidewalk or in the park. For many dog-owners, the regular, albeit fleeting hi-bye moments create a sense of familiarity over time and may eventually trigger a conversation with another dog owner. Conversations would typically begin with exchanging information about their dogs, and after repeated encounters, the conversations may progress to information about the dog-owners themselves.

The regularity and frequency of these casual meetings around the neighbourhood present multiple face-to-face opportunities for people to connect interpersonally and increase their familiarity with one another. Further, shared interests such as children, mutual acquaintances, dogs, and the local neighbourhood provide stimuli for fleeting relations that could eventually become durable friendships.

8. FOOD AND EATING TOGETHER

Shared activity offers people an opportunity to gather and chat. In his conclusion about social cooperation, urban sociologist Richard Sennett (2018, 260) stated that "sociality appears when strangers are doing something productive, together." Food is a cultural ambassador, opening up avenues into different social and cultural worlds. It offers access to new languages, practices, habits, backgrounds, histories, and even worldviews. The act and process of eating a meal together draws people to each other, by offering an external stimulus for sparking conversation – the shared meal provides a "common denominator" as the participants have the opportunity to direct their focus from their differences to their commonalities. For example, a project in Croatia called "A Taste of Home" organizes culinary gatherings where refugees from Iraq, Syria, and Nigeria take turns cooking and introducing their traditional foods, after which they eat together with local Croatians.[17] Through the processes of preparing and serving the food, and then eating it together, participants learn about each other's social and cultural practices. The project provides a platform for interaction and reciprocity between the refugees and the locals, offering a role reversal for them – the refugees assume the role of the host while the Croatians become the guests.

Whyte's (1980) observation of social life in public places underscored how settings that make food and drinks available (or permit them) are used and favoured more than those that do not. Similarly, in Fincher and Iveson's (2008, 159) study of spaces of encounters, they found that libraries and community centres that are inviting, relaxing, and "homey" often have kitchen facilities where people can prepare food and drinks together, and share their meals. In this study of diverse locales, I had a similar observation that public spaces with availability of seating and food broaden the web of opportunities for sociality and conviviality. The arrival of the ice cream cart at Burns Park in the afternoon attracted many children and adults from different ethnicities to experience a shared moment of enthusiasm, breaking the ice between caretakers as their children gathered to enjoy the treats.

Concluding Thoughts: Design for Public Life

Demographic diversification reshapes the relational dynamics between social and cultural groups; these dynamics can undermine and change the possibilities for collective life. In diverse locales, a rethink that is cognizant of the dualistic and dialogical tensions arising from socio-cultural differences between groups is required in the planning and design of spaces for public life. Political philosopher and feminist scholar Iris Marion Young (1990) described her powerful vision of "city life as a normative ideal," inspiring and challenging many urbanists over the decades as societies demographically and culturally diversify. Young's vision challenges cities to be designed for diversity and collective life in a way that retains the spontaneity and freedom of human agency (1990, 238):

> In the ideal of city life, freedom leads to group differentiation, to the forma-tion of affinity groups, but this social and spatial differentiation of groups is without exclusion. The urban ideal expresses difference as a side-by-side particularity neither reducible to identity nor completely other. In this ideal groups do not stand in relations of inclusion and exclusion but over-lap and intermingle without becoming homogeneous.

The findings in this book illustrate that to enable a collective life of inclusion and intermingling in diversity, opportunities for intercul-tural learning must be intentionally planned for in the design of pub-lic environments. This is because cultural boundary crossing processes require the assistance of props, programs, and places to facilitate and catalyse social interaction among people of different social and cultural backgrounds.[18]

Inhabitants in diverse locales grasp the need for intercultural learn-ing but opportunities to engage, it seems, elude them. Similarly, while planners accept the need to plan with diversity in mind, the barri-ers to planning *for* diversity, i.e., to support the building of relations between people of different social and cultural backgrounds, are hard to surmount for a combination of historical, philosophical, political, and practical reasons. Taken together, these findings underscore the need for *co-production* of environments that are conducive to collec-tive intercultural living. The planner cannot do it alone and neither can the inhabitants. Co-production for intercultural living requires reori-entation from the *design of public space* to a *design for public life*, with an awareness of the different kinds of social tensions that sociocultural diversification engender.

8   Conclusions: Conflict and Conviviality in Diversity

Written in the early 1900s, at a time when the United States was undergoing waves of European immigration, Israel Zangwill's play *The Melting Pot*, performed in 1908 on Broadway, envisioned the future of American society to have a common cultural identity that was no longer divided by race, ethnicity, and national origin.[1] America was to be "God's Crucible," according to the protagonist of the play, in which the great fusion of all the different races would produce a composite American identity, "the coming superman" (Glazer and Moynihan [1963] 1970, 289). This popular belief that America melts away racial and cultural differences over time without too much resistance is without doubt being constantly challenged in the first two decades of the twenty-first century. Racial inequity and residential segregation by income, ethnicity, and lifestyle have persisted, while social changes arising from global immigration and technology have created new lines of social differentiation, albeit alongside new possibilities of social connection. The faint hope of the realization of a true melting pot society in America has also affected how countries in Western Europe and East Asia facing demographic diversification confront growing challenges in integrating new immigrants.

This book contributes to the ongoing conversation about coexistence and the corresponding tensions caused by immigration in American society in the past century, and more broadly, about similar issues that have become increasingly relevant for other countries, particularly Western European and East Asian ones, in the last decade. I have deliberately chosen the focus of the study to be on the practical, day-to-day experiences of living in neighbourhoods of sociocultural diversity with different household income levels, with the purpose of offering insights into coexistence as a palpable negotiation between the daily practices of people and the physical environment. Physical environment/space is a medium of the formation of social relations that is active (albeit often unconscious to us) in shaping people's behaviours and the dynamics of interaction between them. The development of human relations and collective life in diversity is affected by how physical space is socially used and shared by inhabitants. This book has shown how and why the physical built environment is critical to the understanding of tensions that emerge in the negotiation of coexistence diversity and in the discussion of efforts towards interculturalism in the following ways:

- by mapping the different social geographies of diverse locales and deriving the socio-spatial dimensions of diverse public environments
- by discussing interculturalism as a daily experience and spatial practice among residents of diverse locales
- by translating the value of intercultural learning and understanding into substantive qualities of public space planning and design

The Diverse Public Realm and Its Promises

Socially and culturally diverse settings present opportunities in and of themselves to increase our capability to relate by *learning* to live with difference, while also presenting challenging conditions that can undermine our capacity for learning. As sociologist Richard Sennett (1970, 138) explained, the ambiguity and uncertainty of diversity in city life stretch a person's mental and emotional development to "deal with dissimilarities," in contrast to the homogeneity, orderliness, and predictability that characterize suburban life, which stagnate human development by limiting an individual's capability to understand and handle differences. Nevertheless, as pointed out by sociologist Lyn Lofland (1973), the volatility of a diverse public realm can be overwhelming and incline us to make simplistic and quick categorical assessments of others. Such assessments, over time, can result in a decline of an individual's capability to relate to others who are unlike himself/herself, which can in turn undermine the individual's capacity to live with differences.

According to political philosopher Martha Nussbaum (2011, 34), having the capability "to live with and toward others, to recognize and show concern for other human beings, to engage in various forms of social interaction, to be able to imagine the situation of another" is one of the ten central capabilities of human development.[2] Capability must not be seen as only limited to an individual's abilities: according to Nussbaum (2011, 20–21), it is a combination of enabling internal and external conditions (political, social, and economic environment) that open up possibilities of choice and freedom for an individual to live a good and dignified life. In other words, an individual's capability to affiliate with or relate to others is an outcome of the person's natural ability as well as the opportunities the person has for connection, learning, and honing of competent relational skills. This highlights the strategic and relational importance of gathering spaces that are freely accessible by different groups in neighbourhoods, if the formation of constructive and equitable intergroup relations is central to the aim of the municipality.

With the growing use of smartphones in public spaces, the public realm seems to be shrinking, as individuals retreat into their private realms of virtual connections, and their ability to be open, alert, and engaged with those who are in their physical proximity becomes severely limited. The use of the smartphone has unwittingly eroded our capability to negotiate socially and culturally diverse relations in person, as we have secured a ready escape from confronting the anxieties arising from being in the presence of difference.

I have underscored the importance of reframing urban planning responsibility from simply the planning/designing *of* public *space* to the planning/designing *for* public *life*. This will require a realignment of urban planning to treat urban space as a medium of human relations rather than as a mere container of functions, and to not shy away from the reformist vision for collective life that characterized urban planning in its beginnings. To enable collective life, cities need not only environments that are welcoming to all social and cultural groups, but also those that can enhance the productivity of social interaction towards forming intercultural relations. In order to plan and design an effective space for intercultural life where dialogical tensions rather than dualistic ones are dominant, dialogue between groups needs to be part of this process from the beginning. In addition, the process requires a good grasp of the intergroup relational dynamics and tensions, user patterns, and spatial design skills, as well as finesse in negotiating the different requirements of city hall. This complexity calls for a necessary co-production of intercultural environments by planners and the city's inhabitants.

Implications beyond Los Angeles

Los Angeles's experience of lived diversity is perhaps ordinary for an immigrant gateway city and typical of cities that are engaged in the international circuit of capital and labour flows. However, given its large population and heterogeneity, together with its sprawling, multicentred and decentralized spatial development, Los Angeles is also paradoxically particular. Thus, to what extent then can we learn from Los Angeles, i.e., what are the implications of this study for cities beyond its borders? I would like to share three deliberations.

First, socio-economic integration is not a precondition for intercultural learning and understanding. One could argue that a matching level of socio-economic status might level unequal power relations and enable the formation of similar lifestyles and values, thereby reducing people's inhibitions to participate in joint activities and engage in intercultural learning. However, as shown in the experience of San Marino's residents, matching levels of socio-economic status and affluence among social and cultural groups do not necessarily dissolve cultural differences or guarantee productive intercultural learning. In the same way, common poverty does not guarantee greater interaction or sharing either, as shown by the struggles residents faced in forming intercultural relations with each other in Central Long Beach. Asian immigrants are seen to have successfully integrated into America because of their high levels of property ownership, education, and household income. However, as the findings of this book have shown, many Asians in the three locales experienced partial integration and selective belonging in America, due to reasons that this book has only lightly touched on.

This brings me to my second point. The desire and struggle of the Asian population to belong is important for America, as the Asian ethnic group, according to American demographer William Frey (2015, 106), will "become a central part of the nation's mainstream in the twenty-first century." With multiple national origins combined with differences in class, education level, and immigration experiences, the Asian ethnicity is a superdiverse category unto itself, making the group's assimilation process, ethnic identification, and characterization varied and challenging to study. America does not face this challenge alone, as many countries in Europe, the Pacific, East Asia, and Southeast Asia are popular destinations for immigrants from Asia. Although often regarded as a "model minority" in many parts of the world due to its hardworking image and potential for upward socio-economic mobility, the Asian immigrant population tends to selectively integrate, preferring to hunker down and focus on the preservation of their roots, languages, traditions, food heritages, and cultures.[3] The continual formation of multiple ethnic towns in Los Angeles, including Little Saigon, Koreatown, Thai Town, Little Persia, Cambodia Town, and Little Bangladesh, attests to the importance that immigrants place on creating a place of familiarity that reminds them of home in a foreign land, and on securing recognition for their children in the future of their adopted country. Given the inclination for cultural territoriality and social fragmentation, the challenge to form collective belonging and opportunities for collective life in multicultural environments cannot be underestimated. It requires sensitive creation and inclusion of accessible opportunities that support the formation of intercultural life within the fabric of ethnic towns.

Finally, it is important to recognize that there is inertia working against interaction with people who look and behave differently from us. I observed a *bundling effect* of social and interpersonal barriers that inhibited intercultural learning from occurring spontaneously. These included negative stereotypes, language barriers, time constraint, and self-sufficiency. When these factors are combined with a lack of community spaces and socio-economic hardship in the neighbourhood, the outcome is the formation of a protective cocoon that shields an individual from the hard work of negotiating of diversity through intercultural learning. The inclination to hunker down with our smartphones can further alienate us from one another as we face greater struggles to have conversations with others.

Through the banal and often silent interpretations of the diverse practices in our environment, our mental constructs of others are profoundly shaped and formed. Our susceptibility to echo chambers and subsequent social exclusion of those with different beliefs, habits, or social and political orientations from our lives can, over time, lead to societal divisions and conflicts. This calls for the work of developing human relations and collective life to be continuous, multifaceted, and multi-locational (in our neighbourhoods, at our schools, and at our workplaces), particularly in socioculturally diverse environments. Like the belayer who diligently varies the slack and friction in the rope to the climber's harness to support the climber's progress and mobility, municipalities will have to recalibrate the social tensions arising from spatial coexistence conscientiously and delicately, if social well-being and economic progress are equally desired.

Appendix 1: Demographic Data

Appendix 1.1. Demographic Data of San Marino in 2010 and 2020

	LA County				San Marino			
	2010	% share of total	2020	% share of total	2010	% share of total	2020	% share of total
Total Population Count	9,818,605		10,014,009		13,147		12,513	
Nativity (*sample estimates)								
TOTAL	9,758,256		10,039,107		13,114		13,194	
Native	6,280,433	64.4	6,637,220	66.1	8,119	61.9	7,664	58.1
Foreign-born	3,477,823	35.6	3,401,887	33.9	4,901	37.4	5,530	41.9
Race/Ethnicity								
TOTAL	9,818,605		10,014,009		13,147		12,513	
White alone	2,728,321	27.8	2,563,609	25.6	4,872	37.1	3,469	27.7
Black or African American alone	815,086	8.3	760,689	7.6	53	0.4	58	0.5
American Indian and Alaska Native alone	18,886	0.2	18,453	0.2	1	0	4	0
Asian alone	1,325,671	13.5	1,474,237	14.7	7,010	53.3	7,581	60.6
Native Hawaiian and Other Pacific Islander alone	22,464	0.2	20,522	0.2	2	0	7	0
Some other race alone	25,367	0.3	58,683	0.6	25	0.2	22	0.2
Two or more races	194,921	2.0	313,053	3.1	329	2.5	484	3.9
Hispanic or Latino	468,7889	47.7	4,804,763	48.0	855	6.5	888	7.1

Language Abilities (*sample estimates)

TOTAL	**9,098,454**		**9,459,860**		**12,601**		**12,729**	
Speak English only	3,966,317	**43.6**	4,086,991	**43.2**	6,561	**52.1**	6,048	**47.5**
Speak another language	5,132,137	**56.4**	5,372,869	**56.8**	6,040	**47.9**	6,681	**52.5**
(and Speak English "not well")	935,460	**10.3**	852,509	**9.0**	613	**4.9**	677	**5.3**
(and Speak English "not at all")	502,802	**5.5**	388,908	**4.1**	227	**1.8**	45	**0.4**
(and Speak English "well" or "very well")	369,3875	**40.6**	413,1452	**43.7**	5,200	**41.3**	5,959	**46.8**
Average Median Household Income (in 2010 and 2019 inflation-adjusted dollars)	**55,476**		**72,797**		**151,800**		**166,607**	

Data source: US Census 2010 and 2020, American Community Survey 5-Year Estimates (2006–10), (2015–19)

Tables from US Census 2010: (P1 SF1), (P5 SF1)

Tables from US Census 2020: (P1), (P2)

Tables from American Community Survey 5-Year Estimates 2006–10: (B05012), (B16005), (B19013), 2015–19: (S1901)

Appendix 1.2. Demographic Data of Central Long Beach in 2010 and 2020

	LA County				Central Long Beach			
	2010	% share of total	2020	% share of total	2010	% share of total	2020	% share of total
TOTAL POPULATION COUNT	**9,818,605**		**10,014,009**		**69,430**		**65,140**	
Nativity (*sample estimates)								
TOTAL	**9,758,256**		**10,039,107**		**69,204**		**66,673**	
Native	6,280,433	**64.4**	6,637,220	**66.1**	41,878	**60.5**	43,032	**64.5**
Foreign-born	3,477,823	**35.6**	3,401,887	**33.9**	27,326	**39.5**	23,641	**35.5**
Race/Ethnicity								
TOTAL	**9,818,605**		**10,014,009**		**69,430**		**65,140**	
White alone	2,728,321	**27.8**	2,563,609	**25.6**	6,352	**9.1**	6,569	**10.1**
Black or African American alone	815,086	**8.3**	760,689	**7.6**	9,909	**14.3**	8,767	**13.5**
American Indian and Alaska Native alone	18,886	**0.2**	18,453	**0.2**	201	**0.3**	192	**0.3**
Asian alone	1,325,671	**13.5**	1,474,237	**14.7**	11,579	**16.7**	9,600	**14.7**
Native Hawaiian and Other Pacific Islander alone	22,464	**0.2**	20,522	**0.2**	345	**0.5**	315	**0.5**
Some other race alone	25,367	**0.3**	58,683	**0.6**	193	**0.3**	331	**0.5**
Two or more races	194,921	**2.0**	313,053	**3.1**	1,419	**2.0**	1,843	**2.8**
Hispanic or Latino	4,687,889	**47.7**	4,804,763	**48.0**	39,432	**56.8**	37,523	**57.6**

Language Abilities (*sample estimates)

TOTAL	**9,098,454**		**9,459,860**		**62,725**		**61,509**	
Speak English only	3,966,317	43.6	4,086,991	43.2	19,154	30.5	21,069	34.3
Speak another language	5,132,137	56.4	5,372,869	56.8	43,571	69.5	40,440	65.7
(and Speak English "not well")	935,460	10.3	852,509	9.0	9,261	14.8	6,806	11.1
(and Speak English "not at all")	502,802	5.5	388,908	4.1	5,884	9.4	3,150	5.1
(and Speak English "well" or "very well")	3,693,875	40.6	4,131,452	43.7	28,426	45.3	30,484	49.5
Average Median Household Income (in 2010 and 2019 inflation-adjusted dollars)	**55,476**		**72,797**		**32,903**		**42,867**	

Data source: US Census 2010 and 2020, American Community Survey 5-Year Estimates (2006–10), (2015–19)
Tables from US Census 2010: (P1 SF1), (P5 SF1)
Tables from US Census 2020: (P1), (P2)
Tables from American Community Survey 5-Year Estimates 2006–10: (B05012), (B16005), (B19013), 2015–19: (S1901)

Appendix 1.3. Demographic Data of Mid-Wilshire in 2010 and 2020

	LA County				Mid-Wilshire			
	2010	% share of total	2020	% share of total	2010	% share of total	2020	% share of total
Total Population Count	**9,818,605**		**10,014,009**		**96,070**		**94,810**	
Nativity (*sample estimates)								
TOTAL	**9,758,256**		**10,039,107**		**100,970**		**95,560**	
Native	6,280,433	**64.4**	6,637,220	**66.1**	41,978	**41.6**	46,948	**49.1**
Foreign-born	3,477,823	**35.6**	3,401,887	**33.9**	58,992	**58.4**	48,612	**50.9**
Race/Ethnicity								
TOTAL	**9,818,605**		**10,014,009**		**96,070**		**94,810**	
White alone	2,728,321	**27.8**	2,563,609	**25.6**	17,694	**18.4**	19,957	**21.1**
Black or African American alone	815,086	**8.3**	760,689	**7.6**	3,863	**4.0**	4,034	**4.3**
American Indian and Alaska Native alone	18,886	**0.2**	18,453	**0.2**	110	**0.1**	128	**0.1**
Asian alone	1,325,671	**13.5**	1,474,237	**14.7**	32,247	**33.6**	31,205	**32.9**
Native Hawaiian and Other Pacific Islander alone	22,464	**0.2**	20,522	**0.2**	48	**0**	39	**0**
Some other race alone	25,367	**0.3**	58,683	**0.6**	337	**0.4**	705	**0.7**
Two or more races	194,921	**2.0**	313,053	**3.1**	1,469	**1.5**	2,713	**2.9**
Hispanic or Latino	4,687,889	**47.7**	4,804,763	**48.0**	40,302	**42.0**	36,029	**38.0**

Language Abilities (*sample estimates)

TOTAL	9,098,454		9,459,860		94,458		89,801	
Speak English only	3,966,317	43.6	4,086,991	43.2	23,002	24.4	27,145	30.2
Speak another language	5,132,137	56.4	5,372,869	56.8	71,456	75.6	62,656	69.8
(and Speak English "not well")	935,460	10.3	852,509	9.0	20,457	21.6	17,611	19.6
(and Speak English "not at all")	502,802	5.5	388,908	4.1	7,840	8.3	6,269	7.0
(and Speak English "well" or "very well")	3693875	40.6	4131452	43.7	43159	45.7	38776	43.2
Average median household income (in 2010 and 2019 inflation-adjusted dollars)	**55,476**		**72,797**		**44,800**		**53,811**	

Data source: US Census 2010 and 2020, American Community Survey 5-Year Estimates (2006–10); (2015–19)

Tables from US Census 2010: (P1 SF1), (P5 SF1)

Tables from US Census 2020: (P1), (P2)

Tables from American Community Survey 5-Year Estimates 2006–10: (B05012), (B16005), (B19013); 2015–19: (S1901)

Appendix 2: List of Participants for the Semi-Structured Interviews in the Three Study Areas

Interview Group refers to:

(1) Residents/users of the neighbourhoods including business owners;
(2) Community organizers who work locally in neighbourhood civic organizations (religious, social services) and city services (park, library, police);
(3) Municipal decision makers (e.g., planners, district representatives) in the city hall.

Appendix 2.1. San Marino

Code	Interview group	Age	Native-born/ immigrant (country of birth)	Ethnicity (description of nationality or ethnicity if given)	Self-identification	Name (replaced with a Pseudonym if given)
SM1	2	not given	not given	White	not given	not given
SM2	3	not given	not given	Latino	not given	not given
SM3	1	30s	native-born	White	not given	not given
SM4	1	40s	native-born	White	not given	not given
SM5	1	70s	native-born	White	not given	not given
SM6	1	70s	native-born	White	not given	not given
SM7	1	not given	immigrant (Hong Kong)	Asian	not given	not given
SM8	1	not given	native-born	White	not given	not given
SM9	1	70s	native-born	White	not given	Dominique Fisher
SM10	1	50s	native-born	White (Croatian, German, Irish)	not given	Mary Philips
SM11	2	50s	native-born	White (American Caucasian)	not given	John Shaw
SM12	3	40s	native-born	White	not given	not given
SM13	3	late 20s	native-born	White-Latino (Mexican, German, Norwegian, Swedish)	not given	not given
SM14	2	40s	native-born	White	not given	not given
SM15	1	late 20s	immigrant (China)	Asian (Chinese)	not given	Zack Shi
SM16	2	not given	not given	White	not given	Luke McDowell

ID	#	Status	Age	Ethnicity	Self-identification	Name
SM17	1	immigrant (Taiwan)	40s	Asian (Chinese from Taiwan)	"I feel like an American. More Asian American second generation."	Naomi Su
SM18	1	immigrant (Taiwan)	not given	Asian	"I am first generation, Asian/Chinese American."	Sandy Cheng
SM19	1	immigrant (Taiwan)	50s	Asian	"I am still Chinese."	Linhui Kao
SM20	1	immigrant (Japan)	40s	Asian	"American Japanese"	not given
SM21	1	native-born	40s	White	not given	Isabelle Anson
SM22	1	immigrant (Taiwan)	60s	Asian	"Chinese living in San Marino, first generation, I am an American citizen."	Lydia Li
SM23	1	native-born	70s	White (Caucasian)	not given	Jennifer Meier
SM24	1	immigrant (Taiwan)	30s	Asian (Taiwanese)	"Taiwanese that is Americanized"	Jonathan Lin
SM25	1	immigrant (Taiwan)	50s	Asian	"Chinese American"	Nick Chang
SM26	1	immigrant (Taiwan)	not given	Asian	not given	not given
SM27	1	immigrant (Taiwan)	not given	Asian	not given	not given
SM28	1	native-born	30s	Asian	"American-born Chinese"	Noah Yu
SM29	1	native-born	30s	Asian	"Officially, I am an American, son of immigrant parents."	Bentley Wang
SM30	1	immigrant (China)	40s	Asian	"An American, a 中国人 (China Chinese)"	Noelle Lu

Appendix 2.2. Central Long Beach

Code	Interview group	Age	Native-born/ immigrant (country of birth)	Ethnicity (description of nationality or ethnicity if given)	Self-identification	Name (replaced with a Pseudonym if given)
LB1	2	not given	not given	White	not given	not given
LB2	2	not given	not given	White	not given	not given
LB3	1	50s	native-born	White (American)	not given	not given
LB4	1	early 20s	native-born	African American	not given	Alina Daniels
LB5	1	50s	immigrant (Cambodia)	Asian (Cambodian, Khmer)	not given	Munny Ly
LB6	1	30s	native-born	African American	not given	John Turner
LB7	1	15–19	native-born	Latino (Mexican)	not given	Joshua Hernandez
LB8	2	50s	immigrant (Cambodia)	Asian (Cambodian)	"Cambodian American, I am a Long Beach resident."	Kosal Sok
LB9	1	15–19	native-born	Latino (Mexican)	"Hispanic, sometimes Mexican"	Laura
LB10	1	50s	native-born	White (homeboy)	"Native, people call me homeboy a lot."	Rich Taylor
LB11	1	30s	native-born	African American/ American/Indian/ Other	"I am XX and trying to survive."	Marteese Owens
LB12	1	early 20s	native-born	Latino	"Mexican American"	Ben Rodriguez
LB13	1	early 20s	native-born	Latino (Mexican American)	"Mexican American"	not given
LB14	2	50s	native-born	African American	not given	not given
LB15	2	60s	native-born	White (Half Sicilian, half-Scot-Irish)	"I am a native from XX (US State)."	Randy Jones

LB16	2	not given	not given	African American	not given	not given
LB17	2	not given	immigrant (Cambodia)	Asian	not given	Chenda So
LB18	2	50s	native-born	Latino	not given	not given
LB19	2	30s	immigrant (Cambodia)	Asian (Cambodia)	not given	not given
LB20	1	15–19	native-born	Latino	"American Mexican"	not given
LB21	2	not given	immigrant (Thailand)	Asian	not given	not given
LB22	2	not given	native-born	White	not given	Kylie Brendon
LB23	2	not given	immigrant (Cambodia)	Asian	not given	not given
LB24	1	50s	immigrant (Vietnam)	Asian (Vietnamese)	"Always Asian"	Anh Dao
LB25	2	late 20s	immigrant (Mexico)	Latino	American	Eric Alvarez
LB26	2	40s	native-born	White (Caucasian)	not given	not given
LB27	3	not given	not given	White	not given	not given
LB28	2	60s	immigrant (Cambodia)	Asian	"Cambodia is my homeland."	Not given
LB29	1	late 20s	native-born	White	"A native from XX city"	Jonathan Anderson
LB30	2	40s	immigrant (Mexico)	Latino (Mexican)	not given	not given
LB31	3	not given	not given	White	not given	not given
LB32	1	15–19	native-born	African American	not given	not given
LB33	1	60s	native-born	Black American	not given	Calvin Jenkins
LB34	3	not given	not given	White	not given	not given

Appendix 2.3. Mid-Wilshire

Code	Interview group	Age	Native-born / immigrant (country of birth)	Ethnicity (description of nationality or ethnicity if given)	Self-identification	Name (replaced with a Pseudonym if given)
MW1	2	not given	not given	White	not given	not given
MW2	1	40s	immigrant (South Korea)	Asian	not given	Mi Young
MW3	1	40s	immigrant (South Korea)	Asian	not given	not given
MW4	1	30s	native-born	White (Irish descent)	not given	not given
MW5	1	40s	immigrant (Mexico)	Latino (Mexico)	not given	not given
MW6	1	30s	native-born	White	not given	not given
MW7	1	40s	native-born	White (US Jewish White)	not given	Courtney Bateman
MW8	1	30s	immigrant (Mexico)	Latino (Mexico)	not given	not given
MW9	1	30s	immigrant (South Korea)	Asian (Korean)	not given	Michael So
MW10	1	70s	native-born	White	not given	Alison Haynes
MW11	1	30s	native-born	African American-Latino	not given	Charlie Brooks
MW12	1	60s	native-born	White (Caucasian)	not given	Jenny Fellow
MW13	1	30s	immigrant (The Philippines)	Asian (Filipino)	"Filipino American because I live in the US but Filipino at heart."	Chloe Castillo
MW14	1	15–19	native-born	Latino (Mexican American)	"More Mexican than Guatemalan."	Luciana Garcia
MW15	1	50s	immigrant (The Philippines)	Asian (Filipino)	"Naturalized immigrant, US Citizen and Filipino."	Not given
MW16	1	30s	immigrant (South Korea)	Asian	"I am Korean."	Yumi Lee
MW17	1	late 20s	native-born	Asian (Filipino)	"Citizen"	Matthew Cruz
MW18	1	not given	not given	White	not given	not given

ID	#	Age	Nativity	Ethnicity	Self-description	Name
MW19	2	60s	native-born	White	not given	not given
MW20	2	30s	immigrant (South Korea)	Asian (Korean)	"Bi-cultural, bilingual, 1.5 generation, America is home. With how I look … Asian American."	Hannah Youn
MW21	2	not given	not given	White	not given	not given
MW22	3	30s	native-born	White (USA)	not given	Marcus Kenny
MW23	2	30s	immigrant (Mexico)	Latino (Mexican Hispanic)	not given	Caleb Torres
MW24	1	60s	native-born	Asian	"Chinese American"	Nancy Lau
MW25	2	not given	not given	White	not given	Kelly Douglass
MW26	1	30s	native-born	African American	"A dreamer, wants to change and grow, woman … I am a Californian."	Tania Johnson
MW27	1	not given	immigrant (Mexico)	Latino	"I am an immigrant."	Not given
MW28	1	15–19	native-born	Latino (Honduran and Black)	not given	Eileen Corez
MW29	1	50s	native-born	African American	"Angleno, part of this community."	Mark Adams
MW30	2	30s	immigrant (El Salvador)	Latino-White-American Indian (Salvadoran)	not given	Damien Torez
MW31	1	50s	native-born	White (American)	not given	Larry Gans
MW32	1	40s	immigrant (South Korea)	Asian (Korean)	"An American because I was educated here, and I know the system."	Liz Joo
MW33	1	early 20s	native-born	Latino (Mexico)	"Mexican descent, semi-American culture."	Lucas Alvarado
MW34	3	not given	not given	African American	not given	not given

Appendix 3: Interview Questions and Survey

Interview Questions

RESIDENTS INTRODUCTION	BUSINESS OWNERS INTRODUCTION
1. Do you live in or near this area/ neighborhood?	1. Do you live in or near this area/ neighborhood? Where do you live?
2. Where do you live? **(STAR sticker)**	2. *How long has your business been operating in this area/city?*
3. Could you please draw the boundary of the neighborhood from your point of view on this map in **yellow**?	3. Could you locate your business with a **(STAR sticker)**?
4. How long have you lived in this area/ neighborhood?	4. Could you please draw the boundary of the neighborhood from your point of view on this map in **yellow**?
5. Where were you born and raised?	5. Where were you born and raised?

INDIVIDUAL ROUTINE NEGOTIATIONS (COGNITIVE MAPPING)	BUSINESS SPACE AS ROUTINE NEGOTIATIONS
I would like to understand about your daily life and social interactions in this neighborhood. Could you tell me a little about this neighborhood and your routine living here?	6. Why did you locate the business here?
	7. Who are your customers? Have they changed?
6. Is there much diversity among the people who live in this neighborhood? Are they concentrated in different areas of the neighborhood? If so, could you mark for me on the map in **Orange** the different concentrations?	8. Are your regulars made up of those who live or work in San Marino? How do you tell?
	9. Can you describe the type of contact your customers have with one another?
7. Are there spaces you avoid? Could you show me in **PINK** where these spaces are? Why?	10. How would you describe your relationship with your customers?
	11. What is the role your business and shop serve in this neighborhood?
8. Where do you go regularly, i.e., at least once a week? Could you mark the places with a **BLUE PEN** and tell me?	12. In your opinion, is this shop used as a place to meet and to get to know neighbors of different cultures and ethnicities living and working in this neighborhood? How so? Why?

(Continued)

(Continued)

INDIVIDUAL ROUTINE NEGOTIATIONS (COGNITIVE MAPPING)	BUSINESS SPACE AS ROUTINE NEGOTIATIONS
a. Who do you meet in these places? b. How do you feel about your encounters with the people in these places? c. Do you talk to the people you meet in these places? What makes it difficult to initiate conversations? Who are the people you talk to? Why? d. What do you talk about? 9. Do you have friends you visit in this neighborhood? Could you mark for me where they live and their ethnicity with a **GREEN PEN**? 10. Who are your neighbors? Where do they come from? 11. How would you describe your relationship between you and your neighbors?	13. Has the neighborhood changed since you started your business here? 14. If so, how has your business changed with the neighborhood? Why? 15. Has it been hard to conduct business in a multi-ethnic setting? 16. How have you catered to a more multi-ethnic crowd? Staffing? Menu?
NEIGHBORHOOD LEVEL NEGOTIATIONS	**NEIGHBORHOOD LEVEL NEGOTIATIONS**
12. Is this neighborhood divided? How so? Why? 13. What issues do people care about and "fight over" in this neighborhood? 14. Does the shared use of neighborhood space cause friction? Which space? What type of use? 15. Has ethnicity/race come up as a point of tension? What about immigrants? 16. Has your identity as an IMMIGRANT/ NATIVE ever been an issue living in this neighborhood? Is it difficult to integrate into this neighborhood? Have you ever felt like an outsider in this neighborhood? Do you feel like you belong in this neighborhood? Elaborate. 17. Did you ever experience discrimination in this neighborhood? If not, do you know of anyone who may have? How do you feel about this?	17. What are business owners here concerned with? What do they "fight over?" 18. Has ethnicity/race come up as a point of tension? What about immigrants? 19. Has your identity as an IMMIGRANT/ NATIVE ever been an issue living in this neighborhood? Is it difficult to integrate into this neighborhood? Have you ever felt like an outsider in this neighborhood? Do you feel like you belong in this neighborhood? Elaborate.

INTERCULTURAL UNDERSTANDING
STATUS AND OPPORTUNITIES

18. Do you think intercultural understanding is lacking in this neighborhood? Why? **(ICU defined as engaging in mutual learning and adaptation between different cultures and ethnicities)**

19. What forms of inter-cultural understanding between neighbors who are ethnically/culturally different can most improve relations in this neighborhood?

20. How important is the neighborhood PARK/LIBRARY/COMMUNITY CENTER as a place for residents living in the neighborhood to meet and get to know each other and their different ethnic cultures? Why do you think so? Could you rank them with 1–3 (1 indicates the best and 3 the least)?

21. Are there other places in or outside the neighborhood that serve this purpose better? Why? Could you mark for me in **RED PEN** the places you think serve this purpose better?

22. How can neighborhood spaces be improved to encourage relationship-building between neighbors of different ethnicities?

INTERCULTURAL UNDERSTANDING
STATUS AND OPPORTUNITIES

20. Do you think intercultural understanding is lacking in this neighborhood? Why? **(ICU defined as engaging in mutual learning and adaptation between different cultures and ethnicities)**

21. What forms of inter-cultural understanding between neighbors who are ethnically/culturally different can most improve relations in this neighborhood?

22. How important is the neighborhood PARK/LIBRARY/COMMUNITY CENTER as a place for residents living in the neighborhood to meet and get to know each other and their different ethnic cultures? Why do you think so? Could you rank them with 1–3 (1 indicates the best and 3 the least)?

23. Are there other places in or outside the neighborhood that serve this purpose better? Why? Could you mark for me in **RED PEN** the places you think serve this purpose better?

24. How can neighborhood spaces be improved to encourage relationship-building between neighbors of different ethnicities?

COMMUNITY ORGANIZERS
INTRODUCTION

1. Do you live in or near this area/ neighborhood?

2. Could you please draw the boundary of the neighborhood from your point of view on this map in **yellow**?

3. How long have you worked lived in this area/neighborhood?

MUNICIPAL OFFICERS
INTRODUCTION

1. Do you live in or near this area/ neighborhood?

2. Could you please draw the boundary of the neighborhood from your point of view on this map in **yellow**?

3. How long have you worked lived in this area/neighborhood?

INSTITUTIONAL ROUTINE
NEGOTIATIONS

4. Is there much diversity among the people who live in this neighborhood? Are they concentrated in different areas of the neighborhood? If so, could you mark for me on the map in **Orange** the different concentrations?

INSTITUTIONAL ROUTINE
NEGOTIATIONS

4. Is there much diversity among the people who live in this neighborhood? Are they concentrated in different areas of the neighborhood? If so, could you mark for me on the map in **Orange** the different concentrations?

(Continued)

(Continued)

INSTITUTIONAL ROUTINE NEGOTIATIONS	INSTITUTIONAL ROUTINE NEGOTIATIONS
5. What would you say are the major issues that concern your organization in this neighborhood? Why? Does the presence of multiple ethnicities and cultures living in proximity create unique challenges for the neighborhood? Why? 6. How important is intercultural understanding as a value and criterion in the plans and policies for this neighborhood? **(ICU defined as engaging in mutual learning and adaptation between different cultures and ethnicities)** 7. How has the organization responded to the presence of different ethnicities and cultures in its programs?	5. What would you say are the major issues that concern municipal decision-makers (like yourself) in this neighborhood? Why? Does the presence of multiple ethnicities and cultures living in proximity create unique challenges for the neighborhood? Why? 6. What are the principles and values that guide the planning of this neighborhood? 7. How important is intercultural understanding as a value and criterion in the plans and policies for this neighborhood? **(ICU defined as engaging in mutual learning and adaptation between different cultures and ethnicities)** 8. How does the city take into account the intercultural interaction and understanding in: a. its conception of plans and policies for the neighborhood? b. its planning and design of neighborhood spaces? 9. Do you plan for neighborhood spaces to be used to encourage intercultural understanding? If you don't and you need to, how would you do so?

NEIGHBORHOOD LEVEL NEGOTIATIONS	NEIGHBORHOOD LEVEL NEGOTIATIONS
8. Is the neighborhood divided? How so? Why? 9. What issues do people living and working in this neighborhood care about and "fight over?" Why? 10. Does the shared use of neighborhood space cause friction? Which space? What type of use? Why? 11. Has ethnicity/race come up as a point of tension? What about immigrants? 12. Have there been any reports of discrimination or hate crimes?	10. Is the neighborhood divided? How so? Why? 11. What issues do people living and working in this neighborhood care about and "fight over?" Why? 12. Does the shared use of neighborhood space cause friction? Which space? What type of use? Why? 13. Has ethnicity/race come up as a point of tension? What about immigrants? 14. Have there been any reports of discrimination or hate crimes?

NEIGHBORHOOD LEVEL NEGOTIATIONS	NEIGHBORHOOD LEVEL NEGOTIATIONS
13. Has identity of IMMIGRANT-NATIVE ever been an issue in this neighborhood? Is it difficult to integrate into this neighborhood? Does the neighborhood face concerns of who belongs and who does not e.g., insider-outsider?	15. Has identity of IMMIGRANT-NATIVE ever been an issue in this neighborhood? Is it difficult to integrate into this neighborhood? Does the neighborhood face concerns of who belongs and who does not e.g., insider-outsider? 16. In your opinion, why do you think this is the case? How may they be resolved? Has the city taken any action toward that end?
INTERCULTURAL UNDERSTANDING STATUS AND OPPORTUNITIES	INTERCULTURAL UNDERSTANDING STATUS AND OPPORTUNITIES
14. Do you think intercultural understanding is lacking in this neighborhood? Why? **(ICU defined as engaging in mutual learning and adaptation between different cultures and ethnicities)** 15. What forms of inter-cultural understanding between neighbors who are ethnic culturally different can most improve relations in this neighborhood? 16. How important is the neighborhood PARK/LIBRARY/COMMUNITY CENTER as a place for residents living in the neighborhood to meet and get to know each other and their different ethnic cultures? Why or why not? Could you rank them with 1–3 (1 indicates the best and 3 the least)? 17. Are there other places in or outside the neighborhood that serve this purpose better? Why? Could you mark for me with a RED PEN the places you think serve this purpose better? 18. How can neighborhood spaces be improved to encourage relationship-building between neighbors of different ethnicities?	17. Do you think intercultural understanding is lacking in this neighborhood? Why? **(ICU defined as engaging in mutual learning and adaptation between different cultures and ethnicities)** 18. What forms of inter-cultural understanding between neighbors who are ethnic culturally different can most improve relations in this neighborhood? 19. How do existing neighborhood spaces (PARK/LIBRARY/COMMUNITY CENTER) support intercultural understanding between neighbors? 20. In order of importance, can you indicate which of these civic spaces serve the goal of intercultural understanding best (1 indicates best, and 3 the least)? 21. Are there other places in or outside the neighborhood that serve this purpose better? Why? Could you mark for me in RED PEN the places you think serve this purpose better? 22. How can neighborhood spaces be improved to encourage relationship-building between neighbors of different ethnicities?

Short Survey on Neighborhood Space

1. **Where in the neighborhood are you likely to start a conversation with a neighbor (a fellow resident of this neighborhood)? Please rank only spaces that apply**

   (1 for most likely, 2 for less, 3 for lesser, 4, 5, 6 ...). You can decide on the range (e.g., 1–3, 1–5, 1–10, 1–25, or even 1–100).

Neighborhood space	Person you recognize/have seen before	Person you are meeting for the first time	Name of place (please state, if possible)
Park			
Library			
Recreation center			
Neighborhood cafes			
Neighborhood restaurants			
Grocery stores			
Church/temples/mosques			
Community events			
Block parties			
Sports events			
Museum/exhibits			
Sidewalks			
Others:			

Now consider a neighbor of another ethnicity/nationality:

2. **Where is a good place to have conversations that can lead to intercultural understanding? Please rank only spaces that apply (1 for most likely, 2 for less, 3 for lesser, 4, 5, 6 ...). You can decide on the range (e.g., 1–3, 1–5, 1–10, 1–25, or even 1–100).**

Neighborhood space	Person you recognize/have seen before	Person you are meeting for the first time	Name of place (please state, if possible)
Invite them home			
Park			

Neighborhood space	Person you recognize/have seen before	Person you are meeting for the first time	Name of place (please state, if possible)
Library			
Recreation center			
Neighborhood cafes			
Neighborhood restaurants			
Grocery stores			
Church/temples/mosques			
Community events			
Block parties			
Sports events			
Museum/exhibits			
Sidewalks			
Others:			

Consider neighbors of different ethnicities/nationalities whom you may recognize but that you do not know:

3. What are the qualities of a public space that can encourage deeper contact and inter-cultural understanding between them?

Qualities	Please rank the qualities that apply (1 for most important)	Description, if any
Safety		
Quiet		
Organized programs for the public		
Spacious		
Accessible spaces		
Common activities		
A Neutral space i.e., not belonging to any groups		
Others (please state):		

4. If you get an opportunity to meet and get to know <u>a neighbor of another ethnicity/nationality</u>, what would you like to know about him/her?

Qualities	Please rank the qualities that apply (1 for most important)	Description, if any
Cuisine		
Philosophy to raise children		
Religious views		
Neighbourly expectations		
Customs		
Background of person (e.g., who they are, what they do, why they are here? etc.)		
Others (please state):		

Demography

Would you mind spending the next 2 minutes answering the questions below? Please check what describes you best.

Your Gender:

Male	
Female	

Your Age (years):

15–19	
20–24	
25–29	
30–39	
40–49	
50–59	
60–69	
70–79	
80 and above	

Education

No schooling	
Nursery to fourth grade	
5th to 10th grade	
11th to 12th grade	
High school graduate	
Some college	
Bachelor's degree	
Master's degree	
Professional school	
Doctorate degree	

How many children do you have?

0	
1–2	
3–5	
6 and more	

How long have you lived in the United States?

Born and raised	
Less than 1 year	
1–5 years	
6–10 years	
11–15 years	
16–20 years	
21 years and more	

Your ethnicity/race and nationality (Check all that apply):

Race categories	Check	Nationality and ethnicity (please state)
Hispanic or Latino		
White		
Black or African		
American Indian and Alaska Native		
Asian		
Native Hawaiian and other Pacific Islanders		
Middle Eastern		
Other ethnicity/race (please state):		

Are you a...

Description	Check	How long have you lived at the current address?
Renter		
Homeowner		

What language/s do you use to interact with your neighbors?

English	
Cantonese	
Chinese	
Hindi	
Khmer	
Korean	
Spanish	
Taiwanese	
Tamil	
Thai	
Vietnamese	
Others:	

.........THANK YOU VERY MUCH FOR YOUR TIME!..........

Would you like to be contacted for future interviews? If so, please leave your name and a way to contact you (either email address or phone number):

Notes

Chapter One

1 See for example, Soja and Scott (1996), Waldinger and Bozorgmehr (1996), Massey and Brodmann (2014), and Okamoto (2014).

2 More than half of the 140 participants completed the mapping component and a total of sixty-eight participants completed the survey.

3 The estimates of the migrant stock were prepared by the Population Division of the Department of Economic and Social Affairs of the United Nations Secretariat. The data presented here refers to the international migrant stock defined as a mid-year estimate of the number of people living in a country or area other than the one in which they were born or, in the absence of such data, the number of people of foreign citizenship. Most statistics used to estimate the international migrant stock were obtained from population censuses, population registers, and nationally representative household surveys (United Nations Population Division 2019, iv).

4 Data sourced from United Nations Population Division (2020).

5 Source: United States Census. n.d. American Community Survey 5-Year Estimates 2006–10. Table B05007: Place of birth by year of entry by citizenship status for the foreign-born population.

6 See for example, Nelson and Tienda (1985), Lopez and Espiritu (1990), Ong, Bonacich, and Cheng (1994), and Okamoto (2014).

7 Based on US Census, the average household median income in Central Long Beach in 2010 is 59.3 per cent of LA County's $55,476. By 2020, the level has slipped slightly to 58.9 per cent (United States Census n.d.).

8 According to discussions on panethnicity by Lopez and Espiritu (1990) and Okamoto (2014), it arises when different ethnic groups see the advantages of adopting a common racial identity, such as Asian American, because of their shared interests, crises, and histories, as well as the fact

that they are identified as homogeneous by outsiders. Panethnicity is used to build and mobilize collective agency to negotiate complex inter-ethnic relations.

9 In a study of the social relations in the super-diverse neighbourhood of Hackney in London, Wessendorf (2014, 11) faced similar difficulties in avoiding the practice of essentializing ethnic categories when analyzing and writing about the relations of many different social and cultural groups in one context. In her book, she decided on an inductive approach that she describes as follows: "I use categorical differentiations and terminologies to speak about people of specific cultural or class background in the way in which my informants use them. I thus write about, for example, 'Vietnamese people' to refer to people's origins, but also as way to refer to my informants' descriptions of others."

10 See for example, Simmel ([1903] 2005), Wirth ([1930] 2005), Allport ([1954] 1979), Milgram (1970), Sennett (1970), Lofland (1973), and Kristeva (1991).

11 Lynch experimented on and developed the method further to uncover how the visual form of a city was imagined by the residents of three American cities, namely Boston, Jersey City, and Los Angeles, and how a city's external physical form interacted with the internal learning process of its residents. He wanted to know how cities could be designed in ways that were coherent (in structure) and aesthetically pleasing (in identity) for urban dwellers. By asking residents of cities to sketch an image of their environment and interviewing them about the image, Lynch wanted not only to develop a method to evaluate the imageability and legibility of the built environment of the city in order to guide better planning and design of cities, but also to highlight the importance of soliciting feedback from residents and users to inform policy and plan-making. The cognitive mapping method, having many applications to understand how human psychology and the built environment interact, quickly became a popular means to research and evaluate a city's spaces, aesthetic qualities, and what a "public image" of a city would look like, if it existed. However, due to its initial small sample size of thirty participants in Boston and its reliance on the ability of participants to draw, Lynch's findings were initially criticized for its limited scientific validity.

12 See Eichenbaum (2015) and Mondschein and Moga (2018).

13 Quoting Valentine (2008, 329), "However, tolerance is a dangerous concept. It is often defined as a positive attitude, yet it is not the same thing as mutual respect. Rather, tolerance conceals an implicit set of power relations. It is a courtesy that a dominant or privileged group has the power to extend to, or withhold from, others. Walzer (1997, 52), for example, writes: 'Toleration is always a relationship of inequality where

the tolerated groups or individuals are cast in an inferior position. To tolerate someone else is an act of power; to be tolerated is an acceptance of weakness.'"

14 See Gleason (1992), Young (1999), Fuchs (1999), Parekh ([2000] 2006), and Kymlicka (2007) for more discussion on multiculturalism.

15 See Entzinger (2000), Abu-Lughod (1991), and Dirlik (2008).

16 Mary Louise Pratt (1991, 34) introduced the term "contact zone" to refer to "social spaces where cultures meet, clash, and grapple with each other, often in contexts of highly asymmetrical relations of power, such as colonialism, slavery, or their aftermaths as they are lived out in many parts of the world today." In Amin's (2002), Wood and Landry's (2008), and Wise and Velayutham's (2009) use of the term, the focus is more about the quotidian common spaces for public sociality in locales where relations of power may not be as highly asymmetrical but nevertheless still exists to the extent of making social interaction across social and cultural differences difficult.

17 Increasingly, writings on multiculturalism and cosmopolitanism have also emphasized the need to encourage purposeful interaction and contact between different cultures in diversity such that the difference between these schools of thought and interculturalism has narrowed. See Taylor (1994), Parekh (1996 and [2000] 2006), Appiah (2006), Sen (2006 and 2007), and Phillips (2007) for more discussion on multiculturalism and cosmopolitanism. A special issue contrasting interculturalism and multiculturalism can be found in the *Journal of Intercultural Studies*, edited by Meer and Modood (2012).

18 See Sandercock (2003), Gudykunst (2004), Jandt (2004), Wood and Landry (2008), and Spencer-Oatey and Franklin (2009) on intercultural communication.

19 The Council of Europe adopts interculturalism as a concept to promote policies and programs in cities that help to manage diversity as an opportunity, advantage, and resource enabler for society through its Intercultural Cities Programme (Council of Europe, n.d.).

20 See Diaz (2005), Cheng (2013), Lung-Amam (2017), and Lee (2019) for discussion of how race and ethnicity shape residential preferences and choices.

21 Baxter and Montgomery (1996).

22 On the night of 3 March 1991, Rodney King, a Black American, was arrested by the police with excessive brutality after a high-speed chase on the 210 Freeway. His brutal arrest was caught on video. The policemen were investigated for misconduct but were subsequently acquitted a year later. This injustice sparked an outbreak of riots, looting, and arson by thousands, including many Black Americans and poor Americans on 29

April 1992. One significant incident that was believed to have fuelled the week-long riots was the fatal shooting of Latasha Harlins in South Central Los Angeles a few days after the brutal arrest of Rodney King in March 1991. Harlins, a Black American teenager, was mistaken for shoplifting by Soon Ja Du, a female Korean American storekeeper. When confronted by the shopkeeper for payment, Harlins and Du got into a scuffle. Thinking that Harlins was reaching into her pocket for a gun, the shopkeeper, out of fear and self-defense, fired a fatal gunshot to the back of Harlin's head. The shopkeeper was found guilty of involuntary manslaughter but the judge ruled against a prison term for the shopkeeper and placed her on probation instead. This sparked a community outcry and escalated the troubled relations between the Black American and Asian American communities in South Central Los Angeles that had been unleashed by two decades of global economic restructuring since the 1970s. According to Ong, Park, and Tong (1994), who started researching the Korean-Black conflict before the 1992 riot, South Central had been suffering from a set of structural conditions that made the area vulnerable to the conflicts arising from cultural differences and prejudices between groups. Throughout the 1970s and 1990s, economic disinvestment by both the private and public sector was apace in South Central due to economic and political restructuring in the United States in response to globalization. Large manufacturing companies such as General Motors, Bethlehem, and Firestone, retailers, and banking services shuttered up, resulting in high levels of joblessness and resident flight from the area. During this time, the area also experienced significant demographic diversification as its residents became 50 per cent Latino and 50 per cent Black, and local Jewish, Chinese, and Japanese merchants sold their businesses to new Korean immigrant proprietors looking for an affordable business location. The poor intercultural communication arising from cultural and linguistic differences between Black and Korean residents further compounded the mistrust between the groups. Black residents saw Korean proprietors who owned many of the liquor stores in the area as exploitative capitalists who made money off the residents but did not reinvest in the community. While Korean American proprietors felt victimized through the suffering of many unaccounted brutal fatalities by Black American gang activities and crimes in the area. These changes and harms piled up over time in South Central, intensifying the intergroup tension that led to the fatal exchange between Harlins and Du. Please see "How the Killing of Latasha Harlins Changed South L.A., Long before Black Lives Matter." *Los Angeles Times*, 18 March 2016, by Angel Jennings.

23 Please see http://www.thedailybeast.com/videos/2012/06/17/rodney -king-can-we-all-get-along.html (accessed 20 March 2017).

Chapter Two

1 This echoes sociologist Saskia Sassen's (1996) question, "Whose city is it?" as economic and cultural globalization recreates new rights to place.

2 A sociological concept introduced by Ferdinand Tönnies in 1887 to conceptualize the societal transition from the values of Gemeinschaft (community) to Gesellschaft (society). These concepts describe a change from durable social ties and personable interaction typical of Gemeinschaft to a formal and rational interaction that characterized Gesellschaft.

3 See Chaskin (1997) for an insightful review of the literature about the neighbourhood and community.

4 See Banerjee, Chakravarty, and Chan (2016).

5 For example, Reinhold, Robert. 1987. "San Marino Journal: East Meets West in Upscale Suburb," *New York Times*. 19 November 1987.

6 Central Long Beach had an annual median household income level of $32,903 in 2010, which was less than 60 per cent of Los Angeles County's level of $55,476, while San Marino's income of $151,800 in 2010 was about 300 per cent of Los Angeles County's level.

7 Familiar strangers, a concept by Stanley Milgram (1972), refers to people whom we do not know personally but only categorically by the visual characteristics and/or the contexts in which they always are associated.

8 "Central Long Beach" is a term that is used to refer to the central geographical location of the area relative to the rest of the City of Long Beach as both a formal official reference and an informal reference by people living in Long Beach to differentiate it from the well-heeled Eastside and Downtown Long Beach. Central Long Beach in this study refers to the census tracts that are enclosed by Redondo Avenue and Long Beach Boulevard on the east–west axis and Pacific Coastal Highway and East Seventh Street along the north-south axis. This is the area where about 69,000 people lived in the 2010 Census and 65,000 people as of the 2020 Census. It includes the area that has the following fifteen census tracts (based on US Census 2010 and 2020): 5754.02, 5753, 5752.01, 5752.02, 5751.01, 5751.02, 5751.03, 5763.01, 5763.02, 5764.01, 5764.02, 5764.03, 5769.01, 5769.03, and 5769.04.

9 San Marino includes the census tracts 4641 and 4642 (based on US Census 2010) and 4641.01, 4641.02, and 4642 (based on US Census 2020). These census tracts represent the entire City of San Marino.

10 "Mid-Wilshire" is a term that has many different spatial extents. Mid-Wilshire in this study refers to the area between Vermont Avenue and La Brea Avenue on the east–west axis (2.5 miles or four kilometres) and Melrose Avenue and Eighth Street on the north-south axis (1.4 miles or two kilometres). It is formed by twenty-seven census tracts,

two zip codes, represented by two neighbourhood councils (Greater Wilshire Neighbourhood Council and the Wilshire Center-Koreatown Neighbourhood Council), three council districts (District 4, 13, 10), at least thirteen neighbourhood associations, cultural districts and business improvement district, not to mention several special planning districts. It includes the following census tracts (based on US Census 2010 and 2020): 1923, 1924.1, 1924.2, 1925.1, 1925.2, 1926.1, 1926.2, 2110, 2112.01, 2112.02, 2113.1, 2113.2, 2114.1, 2114.2, 2115, 2117.01, 2117.03, 2117.04, 2118.02, 2118.03, 2118.04, 2119.1, 2119.21, 2119.22, 2121.01, 2121.02, and 2141.

11 For more examples of the sketch-map method, see Appleyard (1976) and Fenster (2009).

12 The research was cleared by the Institutional Review Board (IRB) as a "NOT Human Subjects Research" on 29 January 2011.

13 There was approximately an equal number of participants and a representation of the major US Census ethnic categories (Whites, Latinos, African Americans, and Asians) in each neighbourhood.

14 Ong (2003, xvi) also had similar experiences in her ethnographic research about Cambodians in Northern California.

15 Of the sixty-eight surveys filled out, fourteen were conducted separately in the different neighbourhood parks (Lacy Park in San Marino, MacArthur Park in Central Long Beach and Burns Park in Mid-Wilshire). I thought it was necessary to increase the number of responses as several interviews did not include the survey because of the time constraints of the participants.

16 See Merriam et al. (2001) for further discussion on the influence of power and positionality doing fieldwork within and across cultures, as well as Henry (2003) for her discussion of her fieldwork experiences as a South Asian diasporic who is constantly posed with contradictory expectations by participants to identify herself in ways that they can place her.

Chapter Three

1 Nicolaides and Zarsadiaz (2017) also described San Marino as a "picturesque enclave" that has been kept that way due to a purposeful design assimilation by its immigrant residents.

2 "Residential Design Guidelines Brochure." *Informational Guide for San Marino Residents*.

3 See Redford (2016) for more discussion of restrictive covenants that governed Los Angeles's urban history.

4 By 2020, according to the US Census, the proportion of Asian residents has grown to 61 per cent of the total population, while the proportion of White residents has fallen to about 28 per cent. In addition, the percentage share

of foreign-born residents has increased from 37.4 per cent in 2010 to 41.9 per cent in 2020 in San Marino. See Appendix 1.1 for details.

5 See Chowkwanyun and Segall (2012), and Abendschein (2012).

6 In a weekly real estate flyer *The View* by Coldwell Bankers in early 2012, it was observed that the luxury property market in Los Angeles had been kept afloat mostly by Chinese investors from China.

7 The concepts of "space of flows" and "space of places" are borrowed from Manuel Castells's ([1996] 2000) discussion of the "network society."

8 San Marino demonstrates what Logan and Molotch (1987) identified as politics of urban space arising from the conflict of use and exchange value by different stakeholders.

9 Mary did not explain what she meant. It could possibly mean Barrio-tract, a stereotypical synonym to refer to the popular, humble, and denser tract-housing district commonly found in Los Angeles where lower-income populations live.

10 The City of San Marino has five elected council members who represent the city for four years. These five members then decide who the Mayor is.

11 See Hamlin, Jessica. 2012. "New Mayor Sun Q&A: Chinese and Non-Chinese Relations." San Marino Patch, 20 March 2012. Accessed 9 February 2022. https://patch.com/california/sanmarino/new-san-marino-mayor-richard-sun-q-a-chinese-and-non-f84d75e05c.

12 Central Long Beach in this study refers to the census tracts that are enclosed by Redondo Avenue and Long Beach Boulevard on the east–west axis and Pacific Coastal Highway and East Seventh Street along the north-south axis. This area has about 69,430 residents. It is politically complex and is represented by four municipal council districts (District 1, 2, 4, and 6).

13 The percentage of foreign-born in the City of Long Beach is about 27.4 per cent and 25.2 per cent in 2010 and 2019 respectively. In Central Long Beach, 39.5 per cent and 35.5 per cent of the population are foreign-born in 2010 and 2019 respectively. The data is sourced from the 2010 and 2019 American Community Survey 5-year estimate.

14 The starkness of the segregated residential landscape and its resulting social distance between groups living in proximity resemble Friedrich Engel's (1845) description of Manchester and Herbert Gans's (1962) description of West End Boston. For a Land Use Plan of Central Long Beach, please visit http://www.longbeach.gov/pages/city-news/long-beach-general-plan-update-is-here/ (accessed 21 August 2018).

15 Source of original map taken from http://longbeach.gov/globalassets/ti/media-library/documents/gis/map-catalog/lb-neighborhoods-with-popest/ (accessed 5 December 2018).

16 Shootings and graffiti tagging have become less frequent but certainly are not absent as gangs continue to work out their differences in public spaces. Active Latino gangs in Central Long Beach are the East Side Longos, West Side Longos, Barrio Pobre, and Barrio Small Town. The Asian gangs are the Asian Boyz and Tiny Rascal Gang. The Black American gangs operating in the area are the Rollin 20s Crips and Insane Crips.

17 "Mid-Wilshire" is used in this book to refer to an approximate area between Vermont Avenue and La Brea Avenue on the east–west axis (2.5 miles or four kilometres) and Melrose Avenue and Eighth Street on the north-south axis (1.4 miles or two kilometres). It is formed by twenty-eight census tracts, has two zip codes, and is represented by two neighbourhood councils (Greater Wilshire Neighborhood Council and the Wilshire Center-Koreatown Neighborhood Council), three council districts (District 4, 13, 10), at least thirteen neighbourhood associations, cultural districts, and business improvement districts, not to mention several special planning districts.

18 According to US Census 2020, census tract 2121.02 in eastern mid-Wilshire has a median household income of $27,167, while census tract 2110 in western mid-Wilshire about two miles (or three kilometres) west along Wilshire Boulevard has a median income of $127,197.

19 In Elijah Anderson's earlier writing *Streetwise: Race, Class, and Change in an Urban Community* (1990), he discussed the concepts of "street etiquette" and "street wisdom" in the context of a biracial neighbourhood located next to a poor Black neighbourhood. These are informal sets of skills that are developed and honed by residents to ensure their personal safety and minimize friction among neighbours in a mixed neighbourhood. Street etiquette is similar to Lofland's (1973) concept of categoric knowing, which uses a generalized and basic set of guidelines to define strangers in order to navigate diversity. Street wisdom, on the other hand is a more dynamic and refined situational competence that enables residents to adapt their responses to the changing street environment in order to navigate with ease among strangers while securing one's personal safety. The "code of the street" (Anderson 1999) is a further refinement of the concept of street wisdom, where individuals grasp how to perform their roles and hold the right social posture to avoid trouble and safeguard their lives in streets capable of physical violence.

Chapter Four

1 Statement Concerning the UNESCO TENSIONS PROJECT (1949–53) https://unesdoc.unesco.org/ark:/48223/pf0000179399 (accessed 14 December 2018).

2 See also Das and Teng's (2000) discussion about how internal tensions between competing factors in strategic alliances can destabilize but also stabilize these business relationships.

3 See Baxter and Montgomery's (1996, 23–31) discussion on Bakhtin's theory of dialogism.

4 See also Sennett's (2012, 18–20) discussion contrasting dialectic and dialogic conversation.

5 https://www.cityofsanmarino.org/government/mayor___city_council _/index.php (accessed 9 February 2022).

6 In the case of homeownership, the house provides the use of a home for its residents, but it also is a space that brings in rent and is a financial asset for the homeowners.

7 See Winton, Richard. 1998. "Stiff Laws Keep San Marino Tidy." *Los Angeles Times*, 1 December 1998, accessed 16 May 2019.

8 Ordinance no. O-18-1343-U. An urgency ordinance of the City of San Marino enacted pursuant to government code section 65858 establishing a moratorium on the issuance of tree removal permits; prohibiting the removal of or damage to certain categories of trees; and declaring the urgency thereof.

9 This is not to say that non-Asian immigrant families do not care as much about good access to educational opportunities. In fact, as Ann Owens (2016, 550) pointed out, households in America care very much about choosing residential options that offer good local school opportunities so that "local school options may be a key mechanism structuring the residential choices of families with children, leading to high income segregation among them than among childless households, for whom school options are less relevant."

10 See Lung-Amam (2017) for her discussion of a similar phenomenon of a white flight from a high school in Fremont, California due to the higher-pressure academic environment brought about by the achievement of Asian immigrant children.

11 I was unable to get conclusive information about the geography of the gang territories in the area between Martin Luther King Jr. Avenue and Cherry Avenue from the participants. Residents highlighted streets that have a presence of gang members. Community organizers mentioned during our interviews that anti-gang interventions are active in the area such as regular removal of graffiti and gang-tagging. Triangulating from these different information sources, I think the area could either belong to multiple gangs that are fighting for control or a truce area because of high scrutiny by the police.

12 Based on their study of African, Middle-Eastern, and European countries, Herbst, McNamee, and Mills (2012) found that "societal fault lines"

of social, economic, and political nature overlap with other forms of differences such as religious inclinations to divide states. Global events exacerbate these fault lines to result in long conflicts and extended violence.

Chapter Five

1 Each block measures about 141 yards (125 metres) by 264 yards (225 metres).
2 In Suzanne Hall's (2012) ethnographic study of a multi-ethnic street in London, she illustrated that interpersonal contact and exchange between diverse individuals along a local commercial street had helped individuals to grow a sense of belonging to the locale and community. Likewise, Wise (2005) found that quality interpersonal and intercultural contact between individuals of different ethnic groups creates a sense of hope and local belonging in a suburb in Sydney. Antonsich (2010) in his study of place-belongingness has also listed personal and social ties as one of the key factors to forming a sense of attachment and belonging in a place.
3 See also Lee (2019) on the discussion of how middle-income Latino and Asian homeowners in Los Angeles felt that while having White neighbours signalled a higher status neighbourhood and better common amenities, some respondents articulated a sense of dis-belonging and of being an outsider.
4 FICO score refers to the credit worthiness computed with the software from Fair Issac Corporation (FICO) that is used by banks and businesses in the United States to measure if a loan should be granted to an individual.
5 According to Altman and Chemers (1980), *primary* territories are the private and intimate spaces of the home that are exclusively owned, controlled, and occupied by a closed group of people. *Secondary* territories are those that regulate access and ownership by group membership, time of use, and even area of use, such as the neighbourhood bar and grocer. *Public* territories are open to the temporal use by any member of the public, for example, a public beach, park, or plaza.

Chapter Six

1 Please refer to chapter 1 for the various definitions of the relational web.
2 According to Lofland (1998, 14), the three realms are defined by the "dominating relational form found in some physical space." In the private realm, the form is "intimate." In the parochial realm, the form is "communal." In the public realm, the form is "stranger or categorical."
3 Mean score computation (for purposes of comparison): The sum of the inverse of the ranking scores is first taken, and the total score in each

category is then divided by the total number of responses in each locale to arrive at the final mean score. 1.0 indicates most important.

4 See Barth ([1969] 1998), Roy (2001), and Appiah (2006).

5 See Alba (1985), Gleason (1992), Taylor (1994), and Schönwälder (2010).

6 As compared to tolerance, "Respect is far more discriminating," according to political theorist Amy Gutmann (1994, 22). Gutmann added, "Although we need not agree with a position to respect it, we must understand it as reflecting a moral point of view … A multicultural society is bound to include a wide range of such respectable moral disagreements, which offers us the opportunity to defend our views before morally serious people with whom we disagree and thereby learn from our differences. In this way, we can make a virtue out of the necessity of our moral disagreements."

Chapter Seven

1 See Chan (2013a), Banerjee, Chakravarty, and Chan (2016), and Chakravarty and Chan (2016) for the discussion of the public environment of new ethnic towns of Southeast Asian diasporas in Los Angeles.

2 In *Good City Form* (1981), Kevin Lynch highlighted that a heterogeneous society would face more difficult decisions in achieving certain important performance dimensions in the planning and design of the city. These dimensions include the following: *fit*, i.e., how well the design of a settlement supports the different behaviours and activities of present and future residents of multiple groups; *access*, i.e., how to ensure equity of reach and connection to resources, people, and services of all social and cultural groups; and *control*, i.e., how to ensure that all residents in diverse settings who are oftentimes in unequal power relations are able to make decisions about the use, repair, and modification of the space.

3 John Friedmann (1987) showed that planning is composed of four major categories of knowledge: policy analysis, social reform, social learning, and social mobilization. These categories have intellectual roots that draw from objective and fact-based sciences that give planning its positivistic elements and subjective and value-based disciplines that form its normative basis.

4 Incorporating diversity into the design of cities has been in discussion for a long time in urban planning scholarship. To name a few early contributions – Jane Jacobs's ([1961] 1989) *The Death and Life of Great American Cities*, Richard Sennett's (1990) *The Conscience of the Eye: The Design and Social Life of Cities*, Lisa Peattie's (1998) *Convivial Cities*, as well as discussions by Zelinksy (1990) and Qadeer (1997) about accommodating multicultural differences in land use zoning, and the writings of

Sandercock (1998, 2000, and 2003), Burayidi (2000), and Umemoto (2001) about greater sensitivity and finesse in planning processes when engaging with cultural differences.

5 See also Thomas Huw (2000), Beebeejuan (2012), Lung-Amam (2017), and Martinez-Ariño et al. (2019).

6 Personal communications in September 2009.

7 Fincher and Iveson (2008) also underscored the importance of "light-touch programming" in facilitating social interaction.

8 Rios's (2015) point made here further echoes Umemoto's (2001) about learning empathy by "walking in another's shoes" and Umemoto and Zambonelli's (2012) facilitating the deliberation of different ways of knowing.

9 See Whyte (1980), Gehl ([1987] 2011), Oldenburg (1989), Fincher and Iveson (2008), and Risbeth and Rogaly (2018).

10 Safety of public spaces is a quality that has been mentioned by, for example, Jacobs ([1961] 1989), Whyte (1980), Lynch (1981), Fincher and Iveson (2008), and Wood and Landry (2008) in their discussions of the characteristics of good urban spaces for living and interaction.

11 For example, Jacobs ([1961] 1989), Oldenburg (1989). and Sennett (1990).

12 Wise (2005, 178), in her study of an Australian neighbourhood, observed that inter-generational and inter-ethnic relations required "some form of ice breaker" in order for relationships to begin. Wise gave the example of how a crisis in an elderly White Australian's family prompted her younger immigrant Asian neighbour to offer help with her daily chores. The crisis broke the ice and established a new relationship between the neighbours.

13 The survey uses an open-ended scale, allowing the respondents to determine the range of the ranking but requiring them to use "1" to indicate the most likely location to have a conversation with a person of another ethnicity. The inverse of the rank is then used to compute the cumulative score for each location. The higher the score, the more likely intercultural learning will occur in that location.

14 See also Gehl ([1987] 2011), Christ et al. (2014).

15 The likelihood values are calculated by taking the score for familiar strangers divided by the value for complete strangers in every category of places. Thus, the Y-axis refers to the ratio between the cumulative score of familiar stranger to complete stranger.

16 See also Whyte (1980), Gehl ([1987] 2011), Shaftoe (2008), and Sennett (2018).

17 Yermi Brenner, "Refugees in Croatia Cook Their Way into Inclusion," *Al Jazeera*, 1 July 2015, http://www.aljazeera.com/indepth/features/2015/06/refugees-croatia-food-syria-nigeria-150624102007686.html.

18 In the recent work of Richard Sennett (2018, 241), *Building and Dwelling: Ethics for the City*, Sennett outlined five elements of the "urban conscious" built environment. He calls for an urban form that "will create the material conditions in which people might thicken and deepen their experience of collective life." These elements are namely, Synchronous, Punctuated, Porous, Incomplete, and Multiple. According to Sennett, a synchronous built environment is one that designs invitations "to mix, rather than impose mixing" (p. 211), and one that minimizes cognitive fragmentation through good orientation of the different activities present. It is also an environment punctuated with places for rest and reflection. The built environment of the open city also has more borders than boundaries where the focus of development is on the porous common edges where different groups intersect rather than in the centre where populations are more homogenous. Further, the built environment is purposefully designed to be incomplete to allow for the emergent and as an open structure that encourages multiplicity instead of "masterminding the whole" and "leaves room for maximum variation and innovation" (p. 237).

Chapter Eight

1 See Glazer and Moynihan ([1963] 1970), Alba (1985), and Gleason (1992).
2 The ten central capabilities are life; bodily health; bodily integrity; ability to use senses, imagination, and thought; ability to have emotions and attachments; practical reason; affiliation; ability to concern for other animals, plants, and nature; ability to play; and ability to have control over one's environment (Nussbaum 2011, 33–5).
3 See Lopez and Espiritu (1990), Zhou and Xiong (2005), Okamoto (2014), and Lee and Zhou (2015).

References

Abendschein, Dan. 2012. "Report: How San Marino Lost Its Whiteness." San Marino Patch. Last modified 27 August 2012. http://sanmarino.patch.com /articles/report-how-san-marino-lost-its-whiteness.

Abu-Lughod, Lila. 1991. "Writing Against Culture." In *Recapturing Anthropology: Working in the Present*, edited by Richard Fox, 137–62. Santa Fe: School of American Research.

Alba. Richard D. 1985. *Italian Americans: Into the Twilight of Ethnicity*. Hoboken, NJ: Prentice Hall.

Allen, James P., and Eugene Turner. 1996. "Ethnic Diversity and Segregation in the New Los Angeles." In *EthniCity: Geographic Perspectives on Ethnic Change in Modern Cities*, edited by Curtis C. Roseman, Hans D. Laux, and Günter Thieme, 1–29. Lanham, MD: Rowman & Littlefield.

Allport, Gordon W. (1954) 1979. *The Nature of Prejudice*. New York: Basic Books.

Altman, Irwin, and Martin M. Chemers. 1980. *Culture and Environment*. Monterey, CA: Wadsworth.

Amin, Ash. 2002. "Ethnicity and the Multicultural City: Living with Diversity." *Environment and Planning A* 34: 959–80.

– 2010. *Cities and the Ethic of Care for the Stranger*. Joint Joseph Rowntree Foundation / University of York Annual Lecture 2010. Accessed 9 February 2022. https://lx.iriss.org.uk/sites/default/files/resources/cities-and-the -stranger-summary.pdf.

Anderson, Elijah. 1990. *Streetwise: Race, Class, and Change in an Urban Community*. Chicago: University of Chicago Press.

– 1999. *Code of the Street*. New York: W.W. Norton.

– 2011. *The Cosmopolitan Canopy*. New York: W.W. Norton.

Angell, Robert C. 1950. "UNESCO and Social Science Research." *American Sociological Review* 15, no. 2 (April): 282–7.

Antonsich, Marco. 2010. "Searching for Belonging – An Analytical Framework." *Geography Compass* 4, no. 6 (June): 644–59.

Appadurai, Arjun. 1990. "Disjuncture and Difference in the Global Cultural Economy." *Public Culture* 2, no. 2 (Spring): 1–24.

– 1996. *Modernity at Large: Cultural Dimensions of Globalization*. Minneapolis: University of Minnesota Press.

Appiah, Anthony. 2006. *Cosmopolitanism: Ethics in a World of Strangers*. New York: W.W. Norton.

Appleyard, Donald. 1976. *Planning a Pluralist City*. Cambridge, MA: MIT Press.

Bailey, F.G. 1996. *The Civility of Indifference: On Domesticating Ethnicity*. Ithaca, NY: Cornell University Press.

Banerjee, Tridib. 2001. "The Future of Public Space: Beyond Invented Streets and Reinvented Places." *Journal of the American Planning Association* 67, no. 1: 9–24.

Banerjee, Tridib. and William C. Baer. 1984. *Beyond the Neighborhood Unit: Residential Environments and Public Policy*. New York: Springer.

Banerjee, Tridib, Surajit Chakravarty, and Felicity H.H. Chan. 2016. "Negotiating the Identity of Diaspora: Ethnoscapes of the Southeast Asian Communities in Los Angeles." In *Space and Pluralism: Can Contemporary Cities Be Places of Tolerance?*, edited by Stephano Moroni and David Weberman. Budapest: Central European University.

Banerjee, Tridib, and Michael Southworth, eds. 1995. *City Sense and City Design: Writings and Projects of Kevin Lynch*. Cambridge, MA: MIT Press.

Barth, Fredrik, ed. (1969) 1998. "Introduction." In *Ethnic Groups and Boundaries: The Social Organization of Culture Difference*, edited by Fredrik Barth, 9–38. Long Grove, IL: Waveland Press.

Baxter, Leslie A., and Barbara M. Montgomery. 1996. *Relating: Dialogues and Dialectics*. New York: Guilford Press.

Beale, Lauren. 2011. "Housing Crisis Hasn't Touched San Marino." *Los Angeles Times*, 31 January 2011. http://articles.latimes.com/2011/jan/31/business/la-fi-san-marino-housing-20110131.

Beck, Ulrich. 2006. *The Cosmopolitan Vision*. Cambridge, UK: Polity Press.

Beebeejuan, Yasminah. 2012. "Including the Excluded: Changing the Understandings of Ethnicity in Contemporary English Planning." *Planning Theory & Practice* 13, no. 4: 529–48.

Blaine, Bruce Evan. 2013. *Understanding the Psychology of Diversity*. 2nd ed. Thousand Oaks, CA: Sage.

Blokland, Talja. 2003. *Urban Bonds: Social Relationships in an Inner City Neighborhood*. Translated by L.K. Mitzman. Cambridge, UK: Polity Press.

Bollens, Scott A. 2006. "Urban Planning and Peace Building." *Progress in Planning* 66, no. 2: 67–139.

Bobo, Lawrence D., Melvin L. Oliver, James H. Johnson Jr., and Abel Valenzuela Jr. 2000. "Analyzing Inequality in Los Angeles." In *Prismatic*

Metropolis: Inequality in Los Angeles, edited by Lawrence D. Bobo, Melvin L. Oliver, James H. Johnson Jr., and Abel Valenzuela Jr., 3–50. New York: Russell Sage Foundation.

Bouchard, Gerard. 2015. *Interculturalism: A View from Quebec*. Translated by Howard Scott. Toronto: University of Toronto Press.

Brenner, Yermi. 2015. "Refugees in Croatia cook their way into inclusion." *Al Jazeera*. 1

July 2015. http://www.aljazeera.com/indepth/features/2015/06/refugees -croatia-food-syria-nigeria-150624102007686.html.

Brubaker, Rogers. 2002. "Ethnicity without Groups." *European Journal of Sociology* 43, no. 2: 163–89.

Brubaker, Rogers, Mara Loveman, and Peter Stamatov. 2004. "Ethnicity as Cognition." *Theory and Society* 33, no. 1 (February): 31–64.

Burayidi, Michael, ed. 2000. *Urban Planning in a Multicultural Society*. Westport, CN: Praeger.

–, ed. 2015. *Cities and the Politics of Difference: Multiculturalism and Diversity in Urban Planning*. Toronto: University of Toronto Press.

California Rich Property Website. Accessed 5 June 2018 (site discontinued).

Cantle, Ted. 2005. *Community Cohesion: A New Framework for Race and Diversity*. New York: Palgrave Macmillan.

Castells, Manuel. (1996) 2000. *The Information Age: Economy, Society, and Culture*. Vol. 1 of *The Rise of the Network Society*. Oxford: Blackwell.

Chakravarty, Surajit, and Felicity H.H. Chan. 2016. ""Imagining and Making Shared Spaces: Multivalent Murals in New Ethnic '-Towns' of Los Angeles." *Space and Culture* 19, no. 4: 406–20.

Chan, Felicity. 2013a. "Spaces of Negotiation and Engagement in Multi-ethnic Ethnoscapes: The 'Cambodia Town' Neighborhood in Central Long Beach, California." In *Transcultural Cities: Border-Crossing and Place-Making*, edited by Jeffrey Hou, 149–63. Oxon: Routledge.

Chan, Felicity. 2013b. "Intercultural Climate and Belonging in the Globalizing-Ethnic Neighborhoods of Los Angeles." *The Open Urban Studies Journal (Special Edition on Public Space and Belonging)* 6: 30–9.

Chaskin, Robert J. 1997. "Perspectives on Neighborhood and Community: A Review of the Literature." *The Social Science Review* 71, no. 4: 521–47.

Chen, Xiangming. 2005. *As Borders Bend: Transnational Spaces on the Pacific Rim*. Oxford: Rowman & Littlefield.

Cheng, Wendy. 2009. "Episodes in the Life of a Place: Regional Racial Formation in Los Angeles's San Gabriel Valley." PhD diss., Department of American Studies and Ethnicity, University of Southern California.

– 2013. *The Changs Next Door to the Diazes: Remapping Race in Southern California*. Minneapolis: University of Minnesota Press.

Chowkwanyun, Merlin, and Jordan Segall. 2012. "How an Exclusive Los Angeles Suburb Lost Its Whiteness." *The Atlantic Cities: Place Matters*, 27 August 2012. http://www.theatlanticcities.com/politics/2012/08/how -exclusive-los-angeles-suburb-lost-its-whiteness/3046/.

Christ, Oliver, Katharina Schmid, Simon Lolliot, Hermann Swart, Dietlind Stolle, Nicole Tausch, Al Ramiah Ananthi, Ulrich Wagner, Steven Vertovec, and Miles Hewstone. 2014. "Contextual Effect of Positive Intergroup Contact on Outgroup Prejudice." *Proceedings of the National Academy of Sciences* 111, no. 11: 3996–4000.

City of Long Beach. n.d. "City of Long Beach Neighborhoods." Accessed 5 December 2018. http://longbeach.gov/globalassets/ti/media-library /documents/gis/map-catalog/lb-neighborhoods-with-popest/.

City of Los Angeles. n.d. "Generalized Land Use." Wilshire LA City Planning. Accessed 4 June 2018. https://planning.lacity.org/plans-policies/community -plan-area/wilshire.

City of San Marino. n.d. City Council Homepage. Accessed 9 February 2022. https://www.cityofsanmarino.org/government/mayor___city_council _/index.php.

City of San Marino. n.d. "Residential Design Guidelines Brochure." Informational Guide for San Marino Residents. Planning and Building Department. Accessed 9 February 2022. https://cms9files.revize.com /sanmarinoca/Residential-Design-Guidelines.pdf.

City of San Marino. 2011. Agenda Report on the "Discussion-Tree Preservation Ordinance." 12 October 2011 (page removed).

City of San Marino. 2012. Agenda Report on the "Discussion-Tree Preservation Ordinance Amendments." 11 January 2012 (page removed).

City of San Marino. 2018. Tree Ordinance no. O-18-1343-U. Accessed 16 May 2019 (page removed).

Council of Europe. n.d. "About Intercultural Cities." Intercultural Cities Programme. Accessed 17 May 2018. https://www.coe.int/en/web /interculturalcities/about.

Coser, Lewis A. 1956. *The Functions of Social Conflict*. New York: Free Press.

Das, T.K., and B.S. Teng. 2000. "Instabilities of Strategic Alliances: An Internal Tensions Perspective. *Organization Science* 11, no. 1: 77–101.

Devadason, Ranji. 2010. "Cosmopolitanism, Geographical Imaginaries and Belonging in North London." *Urban Studies* 47, no. 14 (December): 2945–63.

Diaz, David R. 2005. *Barrio Urbanism: Chicanos, Planning, and American Cities*. New York: Routledge.

Dirlik, Arif. 2008. "Race Talk, Race, and Contemporary Racism." *PMLA* 123, no. 5 (October): 1363–79.

Eichenbaum, Howard. 2015. "The Hippocampus as a Cognitive Map … of Social Space." *Neuron* 87, no. 1: 9–11.

Engels, Friedrich. (1845) 2016. "The Great Towns." In *The Condition of the Working Class in England in 1844*. In *The City Reader*, edited by Richard R. LeGates and Frederic Stout, 6th ed. 53–62. London: Routledge.

Entzinger, Han. 2000. "The Dynamics of Integration Policies: A Multidimensional Model." In *Challenging Immigration and Ethnic Relations Politics: Comparative European Perspectives*, edited by Ruud Koopmans and Paul Statham, 97–118. Oxford: Oxford University Press.

Fainstein, Susan. 2005. "Cities and Diversity: Should We Want It? Can We Plan for It?" *Urban Affairs Review* 41, no. 1: 3–19.

– 2010. *The Just City*. Ithaca, NY: Cornell University Press.

Fenster, Tovi. 2005. "The Right to the Gendered City: Different Formations of Belonging in Everyday Life." *Journal of Gender Studies* 14, no. 3: 217–31.

– 2009. "Cognitive Temporal Mapping: The Three Steps Method in Urban Planning." *Planning, Theory & Practice* 10, no. 4: 479–98.

Fincher, Ruth, and Kurt Iveson. 2008. *Planning and Diversity in the City: Redistribution, Recognition and Encounter*. New York: Palgrave Macmillan.

Fincher, Ruth, and Jane M. Jacobs. 1998. *Cities of Difference*. New York: Guilford Press.

Fischer, Claude S. (1976) 2005. "Theories of Urbanism." In *The Urban Sociology Reader*, edited by Jan Lin and Christopher Mele. London: Routledge.

Frey, William H. 2015. *Diversity Explosion: How New Racial Demographics Are Remaking America*. Washington, DC: Brookings Institution Press.

Friedmann, John. 1987. *Planning in the Public Domain: From Knowledge to Action*. Princeton, NJ: Princeton University Press.

– 2005. "Place-Making as Project? Habitus and Migration in Transnational Cities." In *Habitus: A Sense of Place*, edited by Jean Hillier and Emma Rooksby, 2nd ed. 315–33. Oxon: Routledge.

Fuchs, L.H. 1999. "Race, Religion, Ethnicity and the Civic Culture in the United States." In *The Accommodation of Cultural Diversity*, edited by Crawford Young, 176–211. New York: St Martin's Press.

Gaffikin, Frank, Malachy Mceldowney, and Ken Sterett. 2010. "Creating Shared Public Space in the Contested City: The Role of Urban Design." *Journal of Urban Design* 15, no. 4: 493–513.

Gaffikin, Frank, and Mike Morrissey. 2011. *Planning in Divided Cities: Collaborative Shaping of Contested Space*. Oxford: Blackwell.

Gans, Herbert J. (1962) 1982. *The Urban Villagers: Group and Class in the Life of Italian-Americans*. New York: Free Press.

Gehl, Jan. (1987) 2011. *Life between Buildings: Using Public Space*. Translated by J. Koch. Washington, DC: Island Press.

Gieryn, Thomas F. 2006. "City as Truth-Spot: Laboratories and Field-Sites in Urban Studies." *Social Studies of Science* 36, no. 1: 5–38.

Glazer, Nathan, and Daniel P. Moynihan. (1963) 1970. *Beyond the Melting Pot: The Negroes, Puerto Ricans, Jews, Italians, and Irish of New York City.* Cambridge, MA: MIT Press.

Gleason, P. 1992. *Speaking of Diversity: Language and Ethnicity in Twentieth-Century America.* Baltimore, MD: John Hopkins University Press.

Goetz, Edward G., Rashad A. Williams, and Anthony Damiano. 2020. "Whiteness and Urban Planning." *Journal of the American Planning Association* 86, no. 2: 142–56.

Goffman, Erving. 1963. *Behavior in Public Spaces: Notes on the Social Organization of Gatherings.* New York: Free Press.

Gudykunst, William B. 2004. *Bridging Differences: Effective Intergroup Communication.* 4th ed. Thousand Oaks, CA: Sage.

Gupta, Akhil, and James Ferguson. 1997. "Beyond 'Culture': Space, Identity, and the Politics of Difference." In *Culture, Power, Place: Explorations in Critical Anthropology*, edited by Akhil Gupta and James Ferguson, 33–51. Durham, NC: Duke University Press.

Gutmann, Amy. 2004. "Unity and Diversity in Democratic Multicultural Education: Creative and Destructive Tensions." In *Diversity and Citizenship Education: Global Perspectives*, edited by J.A. Banks, 71–96. San Francisco: Jossey-Bass.

– 1994. "Introduction." In *Multiculturalism: Examining the Politics of Recognition*, edited by Amy Gutmann, 1–24. Princeton, NJ: Princeton University Press.

Hall, Suzanne. 2012. *City, Street and Citizen: The Measure of the Ordinary.* Oxon: Routledge.

Hamlin, Jessica. 2012. "New Mayor Sun Q&A: Chinese and Non-Chinese Relations." San Marino Patch. Video. 20 March 2012. Accessed 9 February 2022. https://patch.com/california/sanmarino/new-san-marino-mayor-richard-sun-q-a-chinese-and-non-f84d75e05c.

Henry, Marsha G. 2003. "'Where Are You Really From?': Representation, Identity and Power in the Fieldwork Experiences of a South Asian Diasporic." *Qualitative Research* 3, no. 2 (August): 229–42.

Herbst, Jeffrey, Terence McNamee, and Greg Mills, eds. 2012. *On the Fault Line: Managing Tensions and Divisions within Societies.* London: Profile Books.

Holston, James, and Arjun Appadurai. 1999. "Introduction: Cities and Citizenship." In *Cities and Citizenship*, edited by James Holston, 1–18. Durham, NC: Duke University Press.

Hudson, Berkley. 1990. "1990s: The Golden Decade: San Marino Portrait of a Community: Blending of Cultures Survives Tensions." *Los Angeles Times*, 15 January 1990. http://articles.latimes.com/1990-01-15/news/ss-100_1_san-marino.

Huw, Thomas. 2000. *Race and Planning: The UK Experience.* London: UCL Press.

Jacobs, Jane. (1961) 1989. *The Death and Life of Great American Cities*: New York: Vintage Books.

Jandt, Fred E. 2004. *An Introduction to Intercultural Communication: Identities in a Global Community*. Thousand Oaks, CA: Sage.

Jang, Mira. 2009. "Koreans and Bangladeshis Vie in Los Angeles District." *New York Times*, 7 April 2009. https://www.nytimes.com/2009/04/07/us/07koreatown.html.

Jennings, Angel. 2016. "How the Killing of Latasha Harlins Changed South L.A., Long before Black Lives Matter." *Los Angeles Times*, 18 March 2016. https://www.latimes.com/local/california/la-me-0318-latasha-harlins-20160318-story.html.

King, Rodney. 2012. "Can We All Get Along?" The Daily Beast. 17 June 2012. Last updated 24 April 2017. http://www.thedailybeast.com/videos/2012/06/17/rodney-king-can-we-all-get-along.html.

Kristeva, Julia. 1991. *Strangers to Ourselves*. Translated by L.S. Roudiez. New York: Columbia University Press.

Krysan, Maria, and Kyle Crowder. 2017. *Cycle of Segregation: Social Processes and Residential Stratification*. New York: Russell Sage Foundation.

Kymlicka, Will. 2007. *Multicultural Odysseys: Navigating the New International Politics of Diversity*. Oxford: Oxford University Press.

Lee, Aujean C. 2019. "'Working towards a Better Future for Ourselves': Neighborhood Choice of Middle-Class Latino and Asian Homeowners in Los Angeles." *Journal of Urban Affairs* 43, no. 7: 941–59. https://doi.org/10.1080/07352166.2019.1657022.

Lee, Jennifer. 2002. *Civility in the City: Blacks, Jews, Koreans in Urban America*. Cambridge, MA: Harvard University Press.

Lee, Jennifer, and Min Zhou. 2015. *The Asian American Achievement Paradox*. New York: Russell Sage Foundation.

Lefebvre, Henri. (1974) 1991. *The Production of Space*. Translated by D. Nicholson-Smith. Hoboken: Wiley-Blackwell.

Li, Wei. 2009. *Ethnoburb: The New Ethnic Community in Urban America*. Honolulu: University of Hawaii Press.

Lofland, Lyn H. 1973. *A World of Strangers: Order and Action in Urban Public Space*. New York: Basic Books.

– 1998. *The Public Realm: Quintessential City Life*. New York: Walter de Gruyter.

Logan, John R., and Harvey L. Molotch. 1987. *Urban Fortunes: The Political Economy of Place*. Berkeley, CA: University of California Press.

Lopez, David, and Yen Espiritu. 1990. "Panethnicity in the United States: A Theoretical Framework." *Ethnic and Racial Studies* 13, no. 2: 198–224.

Loukaitou-Sideris, Anastasia. 1995. "Urban Form and Social Context: Cultural Differentiation in the Uses of Urban Parks." *Journal of Planning Education and Research* 14, no. 2 (January): 89–102.

Lung-Amam, Willow. 2017. *Trespassers? Asian Americans and the Battle for Suburbia*. Berkeley, CA: University of California Press.

Lynch, Kevin. (1960) 1998. *The Image of the City*. Cambridge, MA: MIT Press.

– 1981. *Good City Form*. Cambridge, MA: MIT Press.

– (1985) 1996. "Reconsidering the Image of the City." In *City Sense and City Design: Writings and Projects of Kevin Lynch*, edited by Tridib Banerjee and Michael Southworth, 247–56. Cambridge, MA: MIT Press.

Martinez-Ariño, Julia, Michalis Moutselos, Karen Schönwälder, Christian Jacobs, Maria Schiller, and Alexandre Tandé. 2019. "Why Do some Cities Adopt More Diversity Policies than Others? A Study in France and Germany." *Comparative European Politics* 17 (October): 651–72.

Maslow, Abraham H. 1968. *Toward a Psychology of Being*, 2nd ed. New York: Van Nostrand Reinhold.

Massey, Doreen. 2005. *For Space*. London: Sage.

Massey, Douglas S. 1996. "The Age of Extremes: Concentrated Affluence and Poverty in the Twenty-First Century." *Demography*, 33, no. 4, 395–412.

Massey, Douglas S., and Stefanie Brodmann. 2014. *Spheres of Influence: The Social Ecology of Racial and Class Inequality*. New York: Russell Sage Foundation.

McQuire, Scott, Nikos Papastergiadis, and Sean Cubitt. 2008. "Public Screens and the Transformation of Public Space." *Refractory: A Journal of Entertainment Media* 12, 1–14.

Medina, Jennifer. 2013. "New Suburban Dream Born of Asia and Southern California." *The New York Times*, 28 April 2013. http://www.nytimes.com /2013/04/29/us/asians-now-largest-immigrant-group-in-southern -california.html?pagewanted=all&_r=0.

Meer, Nasar, and Tariq Modood. 2012. "How Does Interculturalism Contrast with Multiculturalism?" *Journal of Intercultural Studies* 33, no. 2: 175–96.

Merriam, Sharan B., Juanita Johnson-Bailey, Ming-Yeh Lee, Youngwha Kee, Gabo Ntseane, and Mazanah Muhamad. 2001. "Power and Positionality: Negotiating Insider/Outsider Status within and across Cultures." *International Journal of Lifelong Education* 20, no. 5: 405–16.

Merry, Sally Engle. 1981. *Urban Danger: Life in a Neighborhood of Strangers*. Philadelphia: Temple University Press.

Milgram, Stanley. 1970. "The Experience of Living in Cities." *Science* 167, no. 3924: 1461–8.

– (1972) 2010. "The Familiar Stranger: An Aspect of Urban Anonymity." In *The Individual in a Social World: Essays and Experiments*, edited by T. Blass. London: Pinter & Martin.

– 1974. "Frozen World of the Familiar Stranger." *Psychology Today* 8: 70–3.

Mondschein, Andrew, and Steve T. Moga. 2018. "New Directions in Cognitive-Environmental Research: Applications to Urban Planning and Design." *Journal of the American Planning Association* 84, no. 3–4: 263–75.

Movoto by Ojo. n.d. Accessed 10 February 2022. https://www.movoto.com /san-marino-ca/.

Nelson, Candace, and Marta Tienda. 1985. "The Structuring of Hispanic Ethnicity: Historical and Contemporary Perspectives." *Ethnic and Racial Studies* 8, no. 1: 49–74.

Ni, Ching-Ching. 2011. "San Marino is Part of Chinese Delegation's Business Tour." *Los Angeles Times*, 29 May 2011. http://articles.latimes.com/2011 /may/29/local/la-me-chinese-investors-20110529.

Nicolaides, Becky M., and James Zarsadiaz. 2017. "Design Assimilation in Suburbia: Asian Americans, Built Landscapes, and Suburban Advantage in Los Angeles's San Gabriel Valley." *Journal of Urban History* 43, no. 2: 332–71.

Nussbaum, Martha C. 2011. *Creating Capabilities: The Human Development Approach*. Cambridge, MA: Belknap Press of Harvard University Press.

Okamoto, Dina G. 2014. *Redefining Race: Asian American Panethnicity and Shifting Ethnic Boundaries*. New York: Russell Sage Foundation.

Oldenburg, Ray. 1989. *The Great Good Place*. New York: Paragon House.

Omi, Michael, and Howard Winant. 2015. *Racial Formation in the United States*. 3rd ed. Oxon: Routledge.

Ong, Aihwa. 2003. *Buddha Is Hiding: Refugees, Citizenship, the New America*. Berkeley, CA: University of California Press.

Ong, Paul, Edna Bonacich, and Lucie Cheng, eds. 1994. *The New Asian Immigration in Los Angeles and Global Restructuring*. Philadelphia: Temple University Press.

Ong, Paul, Kye Young Park, and Yasmin Tong. 1994. "The Korean-Black Conflict and the State." In *The New Asian Immigration in Los Angeles and Global Restructuring*, edited by Paul Ong, Edna Bonacich, and Lucie Cheng. Philadelphia: Temple University Press.

Owens, Ann. 2016. "Inequality in Children's Contexts: Income Segregation of Households with and without Children." *American Sociological Review* 81, no. 9: 549–74.

Parekh, Bhikhu. 1996. "Minority Practices and Principles of Toleration." *International Migration Review* 30, no. 1: 251–84.

– (2000) 2006. *Rethinking Multiculturalism: Cultural Diversity and Political Theory*. New York: Palgrave Macmillan.

Park, Robert E., and Ernest W. Burgess. (1925) 1967. *The City: Suggestions for Investigation of Human Behavior in the Urban Environment*. Chicago: University of Chicago Press.

Peattie, Lisa. 1998. "Convivial Cities." In *Cities for Citizens: Planning and the Rise of Civil Society in a Global Age*, edited by Michael Douglass and John Friedmann, 247–53. Chichester: John Wiley & Sons.

Pettigrew, Thomas F., and Linda R. Tropp. 2006. "A Meta-Analytic Test of Intergroup Contact Theory." *Journal of Personality and Social Psychology* 90, no. 5: 751–83.

– 2011. *When Groups Meet: The Dynamics of Intergroup Contact*. New York: Psychology Press.

Phillips, Anne. 2007. *Multiculturalism without Culture*. Princeton, NJ: Princeton University Press.

Phillips, Susan A. 1999. *Wallbangin': Graffiti and Gangs in L.A.* Chicago: University of Chicago Press.

Pratt, Geraldine. 1998. "Grids of Difference: Place and Identity Formation." In *Cities of Difference*, edited by Jane M. Jacobs and Ruth Fincher, 26–48. New York: Guilford Press.

Pratt, Mary Louise. 1991. "Arts of the Contact Zone." *Profession* (1991): 33–40.

Presner, Todd, David Shepard, and Yoh Kawano. 2014. *HyperCities: Thick Mapping in the Digital Humanities*. Cambridge, MA: Harvard University Press.

Putnam, Robert D. 2007. "E Pluribus Unum: Diversity and Community in the Twenty-First Century: The 2006 Johan Skytte Prize Lecture." *Scandinavian Political Studies*, 30, no. 2: 137–74.

Qadeer, Mohammad A. 1997. "Pluralistic Planning for Multicultural Cities." *Journal of the American Planning Association* 63, no. 4: 481–94.

– 2016. *Multicultural Cities: Toronto, New York, and Los Angeles*. Toronto: University of Toronto Press.

Redford, Laura. 2016. "The Intertwined History of Class and Race Segregation in Los Angeles." *Journal of Planning History* 16, no. 4: 305–22.

Reinhold, Robert. 1987. "San Marino Journal: East Meets West in Upscale Suburb." *New York Times*, 19 November 1987. http://www.nytimes.com/1987/11/19/us/san-marino-journal-east-meets-west-in-upscale-suburb.html.

Rios, Michael. 2015. "Negotiating Culture: Towards Greater Competency in Planning." In *Cities and the Politics of Difference*, edited by Michael Burayidi, 343–61. Toronto: University of Toronto Press.

Risbeth, Clare, and Ben Rogaly. 2018. "Sitting Outside: Conviviality, Self-Care and the Design of Benches in Urban Public Space." *Transactions of the Institute of British Geographers* 43, no. 2 (June): 284–98.

Roy, Ananya. 2001. "'The Reverse Side of the World': Identity, Space, and Power." In *Hybrid Urbanism: On the Identity Discourse and the Built Environment*, edited by Nezar AlSayyad, 229–45. Westport, CT: Praeger.

Rumford, Chris. 2014. *Cosmopolitan Borders*. London: Palgrave Macmillan.

Sandercock, Leonie. 1998. *Towards Cosmopolis: Planning for Multicultural Cities.* Chichester: John Wiley & Sons.

– 2000. "When Strangers Become Neighbours: Managing Cities of Difference." *Planning Theory and Practice* 1, no. 1: 13–30.

– 2003. *Cosmopolis II: Mongrel Cities of the 21st Century.* London: Continuum.

Sassen, Saskia. 1996. "Whose City Is It? Globalization and the Formation of New Claims." *Public Culture* 8, no. 2: 205–23.

Savage, Mike, Gaynor Bagnall, and Brian Longhurst. 2005. *Globalization & Belonging.* London: Sage.

Schmid, Christian. 2008. "Henri-Lefebvre's Theory of the Production of Space: Towards a Three-Dimensional Dialectic." Translated by Bandulasena Goonewardena. In *Space, Difference, Everyday Life: Reading Henri Lefebvre,* edited by Kanishka Goonewardena, Stefan Kipfer, Richard Milgrom, and Christian Schmid, 27–44. Oxon: Routledge.

Schönwälder, Karen. 2010. "Germany: Integration Policy and Pluralism in a Self-Conscious Country of Immigration." In *The Multiculturalism Backlash: European Discourses, Policies and Practices,* edited by Steven Vertovec and Susanne Wessendorf, 152–69. Oxon: Routledge.

Sen, Amartya. 2006. *Identity and Violence.* New York: W.W. Norton.

– 2007. *A Freedom-Based Understanding of Multicultural Commitments.* Talk preceding the Meister Eckhart Award in Cologne on 28 Nov 2007: 1–13. Accessed 9 February 2022. https://www.identity-foundation.de/images/stories/downloads/Vorlesung_Sen_Multikulturalismus_281107.pdf.

Sennett, Richard. 1970. *The Uses of Disorder: Personal Identity and City Life.* New York: W.W. Norton.

– 1990. *The Conscience of the Eye: The Design and Social Life of Cities.* New York: Alfred A. Knopf.

– 2012. *Together: The Rituals, Pleasures and Politics of Cooperation.* New Haven: Yale University Press.

– 2018. *Building and Dwelling: Ethics for the City.* New York: Penguin Books.

Shaftoe, Henry. 2008. *Convivial Urban Spaces: Creating Effective Public Places.* London: Earthscan.

Simmel, Georg. (1903) 2005. "The Metropolis and Mental Life." In *The Urban Sociology Reader,* edited by Jan Lin and Christopher Mele, 23–31. Oxon: Routledge.

Soja, Edward W., and A.J. Scott. 1996. "Introduction to Los Angeles: City and Region." In *The City: Los Angeles and Urban Theory at the End of the Twentieth Century,* edited by Allen. J. Scott and Edward W. Soja, 1–21. Berkeley: University of California Press.

Solis, Miriam. 2020. "Racial Equity in Planning Organizations." *Journal of the American Planning Association* 86, no. 3: 297–303.

Spencer-Oatey, Helen, and Peter Franklin. 2009. *Intercultural Interaction: A Multidisciplinary Approach to Intercultural Communication*. New York: Palgrave Macmillan.

Stea, David. 1974. "Architecture in the Head: Cognitive Mapping." In *Designing for Human Behavior: Architecture and Behavioral Sciences*, edited by Jon Lang, Charles Burnette, Walter Moleski, and David Vachon, 157–168. Strasbourg, PA: Dowden, Hutchinson & Ross.

Steil, Justin, and Laura Humm Delgado. 2019. "Limits of Diversity: Jane Jacobs, the Just City, and Anti-Subordination." *Cities* 91: 39–48.

Suttles, Gerald D. 1972. *The Social Construction of Communities*. Chicago: University of Chicago Press.

Taylor, Charles. 1994. "The Politics of Recognition." In *Multiculturalism: Examining the Politics of Recognition*, edited by Amy Gutmann, 25–73. Princeton, NJ: Princeton University Press.

Tonkiss, Fran. 2013. *Cities by Design: The Social Life of Urban Form*. Cambridge, UK: Polity Press.

Tuan, Yi-Fu. (1977) 2011. *Space and Place: The Perspective of Experience*. Minneapolis: University of Minnesota Press.

Umemoto, Karen. 2001. "Walking in Another's Shoes: Epistemological Challenges in Participatory Planning." *Journal of Planning Education and Research* 21, no. 1: 17–31.

Umemoto, Karen, and Vera Zambonelli. 2012. "Cultural Diversity." In *The Oxford Handbook of Urban Planning*, edited by Randall Crane and Rachel Weber, 197–217. New York: Oxford University Press.

UNESCO Tensions Project (1949-53): Statement. Accessed 14 December 2018. https://unesdoc.unesco.org/ark:/48223/pf0000179399.

United Nations Population Division (Department of Economic and Social Affairs). 2019. International Migration Report 2019. New York: United Nations. Accessed 12 March 2021. https://www.un.org/en/development/desa/population/migration/publications/migrationreport/docs/InternationalMigration2019_Report.pdf.

United Nations Population Division (Department of Economic and Social Affairs). 2020. "International Migration 2020 Highlights." United Nations. Accessed 8 October 2021. https://www.un.org/development/desa/pd/content/international-migrant-stock.

United Nations High Commissioner for Refugees. 2016. "Global Forced Displacement Hits Record High." United Nations. Accessed 30 June 2016. http://www.unhcr.org/en-us/news/latest/2016/6/5763b65a4/global-forced-displacement-hits-record-high.html.

United States Census. n.d. American Community Survey. Accessed 29 September 2021. http://www.census.gov/programs-surveys/acs/.

United States Census. 2010. "Decennial Census of Population and Housing by Decade." Accessed 20 August 2012. http://www.census.gov/2010census/.

United States Census. 2020. "2020 Census." Accessed 28 September 2021. https://www.census.gov/programs-surveys/decennial-census/decade/2020/2020-census-main.html.

Valentine, Gill. 2008. "Living with Difference: Reflections on Geographies of Encounter." *Progress in Human Geography* 32, no. 3: 323–37.

Vertovec, Steven. 2007. "Super-diversity and Its Implications." *Ethnic and Racial Studies* 30, no. 6: 1024–54.

Waldinger, R., and M. Bozorgmehr, eds. 1996. *Ethnic Los Angeles*. New York: Russell Sage Foundation.

Walzer, Michael. 1995. "Pleasures and Costs of Urbanity." In *Metropolis: Center and Symbol of Our Times*, edited by Philip Kasinitz. New York: New York University Press.

Watt, Paul. 2009. "Living in an Oasis: Middle-Class Disaffiliation and Selective Belonging in an English Suburb." *Environment and Planning A: Economy and Space* 41, no. 12: 2874–92.

Wessendorf, Susanne. 2014. *Commonplace Diversity: Social Relations in a Super-diverse Context*. London: Palgrave Macmillan.

Whyte, William H. 1980. *The Social Life of Small Urban Spaces*. Washington, DC: Conservation Foundation.

Williams, Rashad Akeem. 2020. "From Racial to Reparative Planning: Confronting the White Side of Planning." *Journal of Planning Education and Research* (August 2020). https://doi.org/10.1177/0739456X20946416.

Winton, Richard. 1998. "Stiff Laws Keep San Marino Tidy." *Los Angeles Times*, 1 December 1998.

Wirth, Louis. (1930) 2005. "Urbanism as a Way of Life." In *The Urban Sociology Reader*, edited by Jan Lin and Christopher Mele, 32–41. Oxon: Routledge.

– 1949. "Comments on the Resolution of the Economic and Social Council on the Prevention of Discrimination and the Protection of Minorities." UNESCO *International Social Science Bulletin* 1, no. 3–4: 137–54.

Wise, Amanda. 2005. "Hope and Belonging in a Multicultural Suburb." *Journal of Intercultural Studies* 26, no. 1–2: 171–86.

Wise, Amanda, and Selvaraj Velayutham. 2009. "Introduction: Multiculturalism and Everyday Life." In *Everyday Multiculturalism*, edited by Amanda Wise and Selvaraj Velayutham, 1–17. New York: Palgrave Macmillan.

Wood, Phil, and Charles Landry. 2008. *The Intercultural City: Planning for Diversity Advantage*. London: Earthscan.

Young, Crawford. 1999. "Case-Studies in Cultural Diversity and Public Policy: Comparative Reflections." In *The Accommodation of Cultural Diversity*, edited by Crawford Young, 1–18, New York: St. Martin's Press.

Young, Iris M. 1990. *Justice and the Politics of Differences*. Princeton, NJ: Princeton University Press.

Yuval-Davis, Nira. 2006. "Belonging and the Politics of Belonging." *Patterns of Prejudice* 40, no. 3: 197–214.

Zajonc, R.B. 2001. "Mere Exposure: A Gateway to the Subliminal." *Current Directions in Psychological Science* 10, no. 6: 224–8.

Zelinsky, W. 1990. "Seeing Beyond the Dominant Culture." *Places* 7, no. 1: 33–5.

Zhou, Min, and Yang Sao Xiong. 2005. "The Multifaceted American Experiences of the Children of Asian Immigrants: Lessons for Segmented Assimilation." *Ethnic and Racial Studies* 28, no. 6: 1119–52.

Index